Tae Young Han

Phased Array Ultrasonic Testing of Dissimilar Metal Welds using Geometric based Referencing Delay Law Technique

TUD*press*

Tae Young Han

Phased Array Ultrasonic Testing of Dissimilar Metal Welds using Geometric based Referencing Delay Law Technique

TUDpress
2014

Die vorliegende Arbeit wurde am 20. Dezember 2013 an der Fakultät Elektrotechnik und Informationstechnik der Technischen Universität Dresden als Dissertation eingereicht und am 27. März 2014 verteidigt.

Vorsitzender der Promotionskommission:
Herr Prof. Dr.-Ing. habil. Dr. h.c. K.-J. Wolter

Gutachter:
Herr Prof. Dr.-Ing. habil. N. Meyendorf
Herr Prof. Dr. Jong-Bum Seo

Bibliografische Information der Deutschen Nationalbibliothek
Die Deutsche Nationalbibliothek verzeichnet diese Publikation in der Deutschen Nationalbibliografie; detaillierte bibliografische Daten sind im Internet über http://dnb.d-nb.de abrufbar.

Bibliographic information published by the Deutsche Nationalbibliothek
The Deutsche Nationalbibliothek lists this publication in the Deutsche Nationalbibliografie; detailed bibliographic data are available in the Internet at http://dnb.d-nb.de.

ISBN 978-3-944331-62-1

Verlag der Wissenschaften GmbH
Bergstr. 70 | D-01069 Dresden
Tel.: 0351/47 96 97 20 | Fax: 0351/47 96 08 19
http://www.tudpress.de

Gesetzt vom Autor.
Printed in Germany.

Technische Universität Dresden

Phased Array Ultrasonic Testing of Dissimilar Metal Welds using Geometric based Referencing Delay Law Technique

Tae Young Han

von der Fakultät Elektrotechnik und Informationstechnik
der Technischen Universität Dresden

zur Erlangung des akademischen Grades eines

Doktoringenieurs

(Dr.-Ing.)

genehmigte Dissertation

Vorsitzender: Herr Prof. Dr.-Ing. habil. Dr. h.c. K.- J. Wolter

Gutachter: Herr Prof. Dr.-Ing. habil. N. Meyendorf
Herr Prof. Dr. Jong-Bum Seo

Tag der Einreichung: 20.12.2013
Tag der Verteidigung: 27.03.2014

Contents

Chapter1. Introduction

The use of austenitic materials is demanded both by the mechanical and by high temperature or corrosive circumstances in nuclear power plants. The term 'austenitic' refers to a metallurgical phase of iron in which it has the face-centered cubic (FCC) crystal structure. Steel exists in its austenitic (or γ-) phase at the temperatures of around 1000°C.

The designing dissimilar metal weld (DMW) using austenitic materials, in particular, is well adopted due to its' higher sustainability in nuclear power plants [1]. The term of dissimilar metal welds is in common use within the nuclear industries and refers to cases where the joint is formed between a low-alloy ferritic steel and an austenitic stainless steel.

There are two representative types of DMW used in nuclear reactors under operation world-wide. One is the Westinghouse type and the other is Combustion Engineering (CE) type. The prototypical locations of individual DMWs are shown in Figure 1.1.

There are several main joint configurations used in the nuclear power plants which contain weld deposits that are in the austenitic state at operating temperatures. These include:

I. the safe-end welds, which connect the nuclear pressure vessel to its primary cooling circuit,

II. subsea joints on hubs/manifolds, the use of which is rapidly increasing with the need to access oil/gas fields in deep waters,

III. corrosion resistant alloy (CRA) claddings at flow lines which require the use of austenitic weld fillers in their girth welds due to metallurgical reasons.

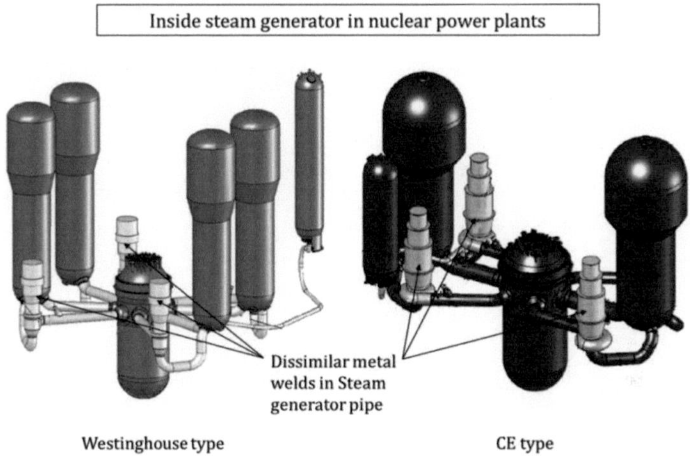

Figure 1.1: Main location of dissimilar metal welds in nuclear power plants- Westinghouse type (left) and CE type (right) (by the courtesy of Korea Hydro and Nuclear Power R&D Institute)

A particular feature of some DMW's is the use of buttering between the ferritic component and the weld deposit, which is deposited before the weld and often from a different weld position. Figure 1.2 shows a common design of the joint with DMW's.

Generally, high nickel content in stainless steel retains austenitic phase on cooling to room temperatures. The fast cooling rates and channels yield what is termed dendritic epitaxial grain growth over successive weld deposits leading to the coarse and elongated grain structure characteristic of austenitic welds [2]. The solidified weld with the FCC structure shows strong elastic anisotropy, i.e. it indicates differing elastic coefficient for different grain orientations in which it is stressed [3].

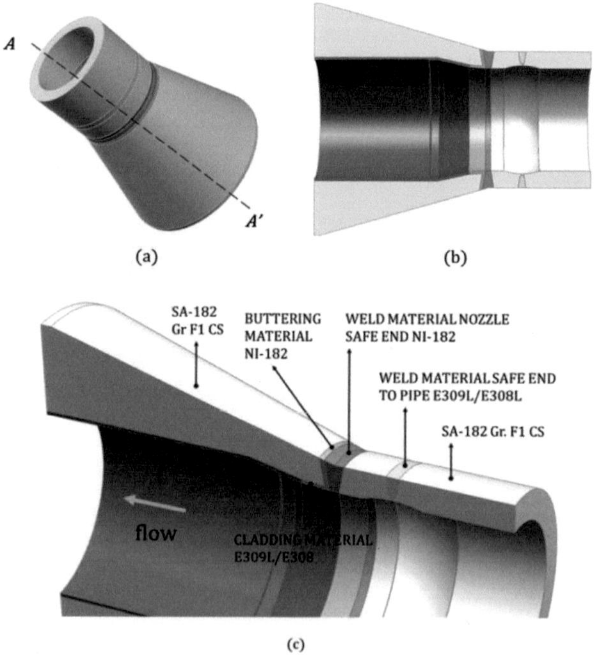

Figure 1.2: Dissimilar metal weld in steam generator in nuclear power plants (a) ISO view, (b) A-A′ Section view and (c) Typical composition of dissimilar metal weld

For these reasons, the propagation of ultrasound waves through the austenitic material is intricate due to its inhomogeneous and anisotropic nature, even though non-destructive testing (NDT) using ultrasound has been widely used for crack detection in austenitic welds. The major obstacles for the sound wave propagation to be uniform are the coarse size of grains originated during solidification of the weld deposits and the differential texture, i.e. anisotropy present across the grain boundaries [4, 5]. Such a characteristic yields considerable limitations as follows:

I. decreased signal-to-noise ratio (SNR) or increased backscattered energy, primarily from the grain boundaries which exhibit significant differences in texture across them; this energy increases the 'noise' in the received ultrasonic signals [6, 7],

II. increased attenuation of the propagating sound waves, limiting the range of inspection [8, 9, 10],

III. increased distortion in the propagation of sound waves, including what is termed 'beam skewing'. For example, whereas in the isotropic ferritic material the wave front would propagate uniformly, in the anisotropic austenitic material this is no longer the case [11, 12].

To summarize, during inspection of austenitic welds, particularly in thick section welds it can be difficult to locate the defect. Due to the problems descripted above, it becomes often the controversial issue for NDT engineers that the individual differences of skill are inevitable which is also degrading reliability of testing results. Particularly, compared to the other NDT methods, the results of ultrasonic testing depend on individual skills, types of equipment and procedures. M. Beth and J. L. Monjaret [13] reported the problematic points based on the results of round robin tests for locating inter-granular stress corrosion cracking (IGSCC) signals in pipe lines in the nuclear power plants. It was found that there were problems not only with capability of locating the cracks but also with the reliability of the ultrasonic testing during pre- and in-service inspection [14]. Hence, in order to improve the reliability for the testing results, the need for performance demonstration (PD) of DMW's for ultrasonic testing system in accordance with the ASME Code, Section XI [15] was brought to the fore. For the PD, mockups or reference blocks with the artificially induced defects which are similar to the real flaws being generated are produced using the same material and dimension. At the PD, ultrasonic testing inspectors should detect the defects according to specified procedure and device and should be qualified. The origin of PD system comes from the research outcomes of "Program for the Inspection of Steel Components (PISC)", which started in Europe in 1974, the finding of IGSCC in piping system at American boiling water reactor nuclear power plants in the early 1980s, and the results of the round robin test conducted at the Electric Power Research Institute (EPRI) [16]. B. Yoon et. al [17] reported the comparative study between the PD and the traditional dB-drop method for finding artificial flaws in DMW specimen through a round robin test participating seven skilled ultrasonic testing inspectors. According to the requirements of ASME Code Section XI Appendix VIII, the defect length sizing must be within or below the root mean square error (RMSE) of 0.75" (19.05 mm) and the depth sizing must be within or below the RMSE of 0.125" (3.175 mm). The test results, in particular in the case of the depth sizing, showed that PD system still has the remarkable deviation of the individual skills independent on both the technical experience and the qualification level. B. Yoon et. al reported that the main reason of the deviation in depth sizing would be difficulty of identifying crack tips due to beam skew and insufficient sound energy to be reflected. Figure 1.3 shows the round robin test results of the defect length sizing and the depth sizing.

Thanks to those studies, the use of reference blocks has standardized as a vehicle to improve the field measurements of ultrasonic testing. For example, a reference block should be employed in testing procedures to setup the required parameters and to decide the threshold of signal amplitudes [15].

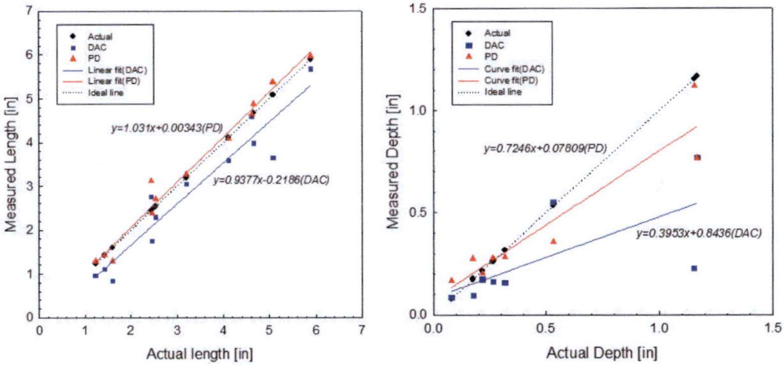

Figure 1.3: The deviation of length sizing (left) and depth sizing (right) of defects by dB-drop and performance demonstration method which were performed by seven ultrasound testing inspectors, DAC is distance-amplitude compensation corresponding to dB-drop method and PD is performance demonstration method [17]

Hence, the main aim of this dissertation is to overcome the individual skill deviation and to provide a new practical method of not only finding flaws including tip signals but also locating them in DMW's by using a reference block with artificial flaws based on a phased array ultrasonic testing combined with full matrix capture (FMC) and total focusing method (TFM).

The reference block was produced from the DMW's mockup which is used for PD of NDT engineers in nuclear power plants. The specimen was geometrically investigated and the DMW's region was analyzed by metallography, grazing incidence ultrasound microscopy (GIUM) and electron backscatter diffraction (EBSD) technique. For the validation of this study, two notches were produced on the bottom side by electrical discharge machining (EDM). One notch is located in a transversely isotropic region in front of the austenitic weld and the other is placed behind the weld.

To help understanding the wave propagation in the weld region before the measurements, the wave propagation was visualized by scanning laser vibrometry (SLV). In addition, an austenitic calibration block, so called K1, was demonstrated to formulate a hypothesis.

In the measurements, both a phased array by sectorial scan mode and FMC were employed for data acquisition. In order to reconstruct the images, TFM for calculating focal laws was used. For sound velocity measurements in vertical and horizontal axes of the specimen, synthetic aperture focusing technique (SAFT) was used.

Thanks to the great advantage of FMC and TFM, a geometry-based image processing using the reference block can be used for the practical quantitative ultrasound testing. The first step is to obtain a basic-data-set of sound velocity values in horizontal and vertical axes obtained from the EDM notch where the waves pass into the transversely isotropic region by the FMC-TFM combination technique. When the sound velocity values were directly obtained, the transversely isotropic region became a reference and was then used to determine the focal laws of the EDM notch behind the austenitic weld by the geometry-based computation. The calculated focal laws are then used for the phased array sectorial scan mode for validation of the approach.

This technique is called as Referencing Delay Law Technique (RDLT). The RDLT can provide a higher probability of detection (POD) even locating the notch tip diffraction signals and overcome the degradation of the testing results due to the wave distortion by the austenitic weld. In addition, the technique can also minimize the individual difference in acquisition of testing data because it uses a full data set covering the whole region of interest (ROI) and focusing at any point in the grid of the region.

Chapter 2.

Principle of phased array ultrasonic testing

2.1 Introduction

The conventional ultrasonic transducers consist of only one element that has a function both of transmitter and receiver. For beam steering around a desired angle, the transducer is typically mounted on an acryl wedge so that the beam can propagate with the required mean incident angle into the test specimen.

While conventional ultrasonic testing methods provide a single element transducer achieved by manual scanning with different angles, phased array transducers consist of elements in array which are pulsed at different time delay and this enables the beam at the required angle to be transmitted. Transmitting and receiving the sound waves depend on the pulse sequence which can be calculated by focal law. The physical principle to generate sound wave is therefore identical to a conventional ultrasonic beam [18, 19].

In other words, phased array is simply considered as a method of both transmitting and receiving ultrasonic waves by foaming the wave front. The advantages of using phased array, however, are manifold in comparison to the conventional technique. The scanning methods of the phased array, e.g., either sector or the azimuthal scan, are capable of electronically controlling the beam angle. Hence it is possible to sweep over an angular range by which it is capable to rapidly scan the specimen just holding the transducer without any manual movement.

The other remarkable advantage is the capability of beam focusing especially when a thin test piece should be inspected. This provides the ultrasound wave with a desired depth of focus. The unavoidable limitation of phased array is that the imaging resolution is mainly determined by the pitch, e.g. distance between center-to-center elements, which is limited by half the wavelength and causes the problem with grating lobes [19].

2.2 Fundamentals of ultrasound

The terms of sound and ultrasound are typically used to explain the propagation of a mechanical vibration in different frequency ranges. Ultrasound corresponds to a mechanical wave whose frequency is above 20 kHz. A time-varying deformation or vibration in materials is a basis of ultrasonic testing method. Every material is comprised of atoms which can be forced into vibrational motion according to its equilibrium positions. When no stress or tension is given to a material, the individual particles behave elastic oscillations. When the medium particles are displaced from their equilibrium positions, subsequently internal restoring forces occur. Such elastic restoring forces between particles result in the oscillatory motions of the medium.

2.2.1 Wavelength, frequency and velocity

The wavelength is directly proportional to the wave propagation velocity and inversely proportional to the wave frequency. Wave length λ is given by the following equation:

$$\lambda = \frac{c}{f} = ct \qquad (2.1)$$

where c is the wave propagation velocity and f is frequency and t is period of beam propagation. As the wavelength becomes shorter, the probability of detection (POD) of smaller fine discontinuities is increased [20]. In order to select the proper frequency for experiment, the grain structure and thickness must be considered. As frequency increases, a scattering effect from the course grain structures increases [21], for example, cast materials.

2.2.2 Wave propagation

The sound waves propagate due to the vibrations or oscillatory motions of particles in the medium. A motion of particles in nearest neighbor affects an individual particle and the particle is influenced both by inertial and by elastic restoring.

2.2.3 Wave velocity, elastic constants, and scattering

Since the density and the elastic constants of the material are different for different materials, sound wave propagates at different velocities in different materials. The relationship between wave velocity, elastic modulus, and density is given by the following equation:

$$c^2 = \frac{M}{\rho} \qquad (2.2)$$

where c^2 is the wave velocity, M is the relevant elastic modulus, and ρ is the density of the material. The elastic constants in isotropic materials are the same for all directions in the material but in anisotropic materials the elastic constants are different for each direction.

As sound waves propagate into the material, the amplitudes of the waves decrease with distance. One of the main reasons of this weakening is scattering. When density and wave velocity of the material change, these result in scattering at the interfaces [22]. The transmission coefficient and the reflection coefficient of an interface are typically expressed by T and R, respectively. Then both coefficients (for a planar discontinuity case) can be given by the acoustic impedance z, before the planar discontinuity, $z_1 = \rho_1 c_1$ and after the planar discontinuity $z_2 = \rho_2 c_2$, where ρ_1 and ρ_2 are the densities and c_1 and c_2 are the respective wave velocities. When p_i is denoted as initial pressure, the pressure of the reflected waves, p_r, is given by:

$$p_r = Rp_i = \frac{z_2 - z_1}{z_2 + z_1} p_i \tag{2.3}$$

and the pressure of the transmitted waves, p_t, is given by

$$p_t = Tp_i = \frac{2z_2}{z_2 + z_1} p_i \tag{2.4}$$

2.2.4 Refraction, Snell's law and mode conversion

When ultrasound waves pass through the interface between two media which have different acoustic impedances at a certain angle, both reflected and refracted waves are generated. Snell's Law states the relationship between the sound velocity and the beam angle as given by the following equation and illustrated by the Figure 2.1.

$$\frac{\sin\theta_1}{c_{P_1}} = \frac{\sin\theta_2}{c_{P_2}} \tag{2.5}$$

where c_{P_1} is the P-wave velocity in medium 1, c_{P_2} is the P-wave in medium 2. $c_{P_1'}$ is the reflected wave at the same angle as the incident sound wave because the two waves passes through the same medium with the same sound velocity.

When the difference of sound velocities between two materials becomes bigger, the refracted angle becomes larger as well. Compared to the longitudinal wave, the angle of refraction of the shear wave is smaller because shear waves pass relatively slower than longitudinal waves.

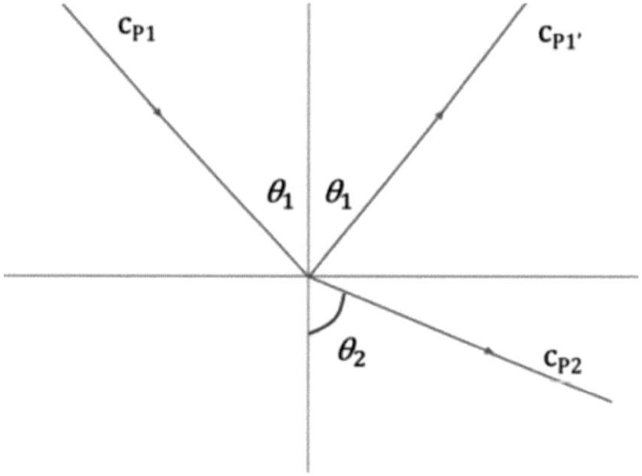

Figure 2.1: A schematic image of Snell's law

This effect caused by the difference of the sound velocity affects the reflection occurred at the interface. Mode conversion can also be described by Snell's Law as given by the following equation and Figure 2.2 shows the schematic pattern of the waves at reflection and at refraction:

$$\frac{\sin\theta_1}{c_{P_1}} = \frac{\sin\theta_2}{c_{P_2}} = \frac{\sin\theta_3}{c_{S_1}} = \frac{\sin\theta_4}{c_{S_2}} \tag{2.6}$$

where c_{P_1}, c_{P_2} and c_{S_1}, c_{S_2} are the P-wave and S-wave velocities in medium 1 and 2, respectively.

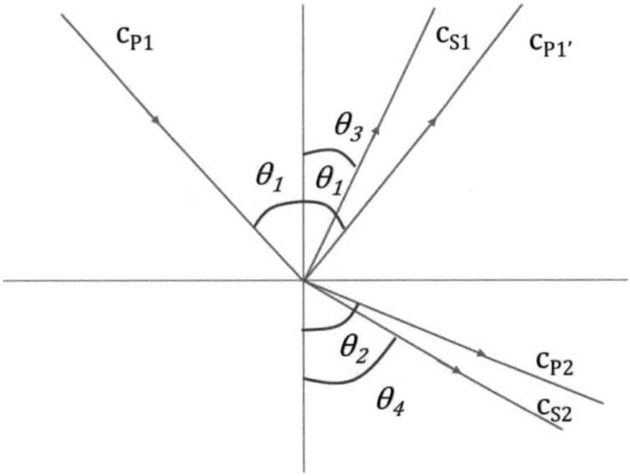

Figure 2.2: A schematic image of Snell's Law including mode conversion

When a sound wave hits an interface between different materials and the incident angle is not normal to the interface, mode conversion occurs. When a P-wave encounters an interface at a certain angle, for example, some of the sound pressure can cause particle movement in the transverse direction to initiate a transverse wave. Figure 2.3 illustrates the effects by a 2-D simulation based on the elastodynamic finite integration technique (EFIT) [23].

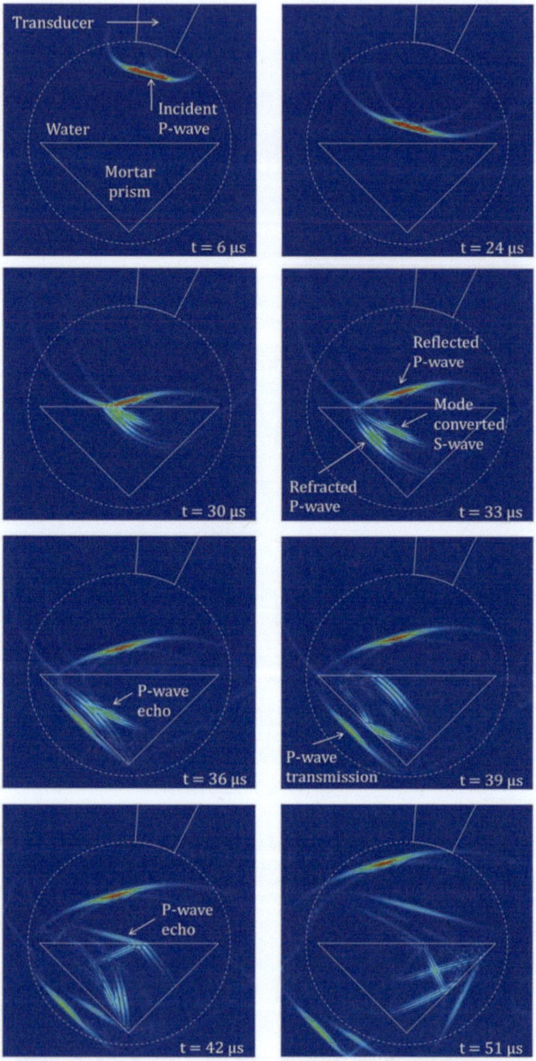

Figure 2.3: Simulated mode conversion within prism-shaped material sample surrounded by water by 2-D EFIT (courtesy of Dr. Frank Schubert, Fraunhofer IZFP-D)

2.2.5 Imaging modes

An A-scan (amplitude scan) represents the amount of reflected pulse amplitude against time of flight (TOF). The relative amount of received sound pressure is displayed along the vertical axis and the elapsed time is represented along the horizontal axis. A-scan images can be displayed as bipolar rectified signals or radio frequency signals as shown in Figure 2.4.

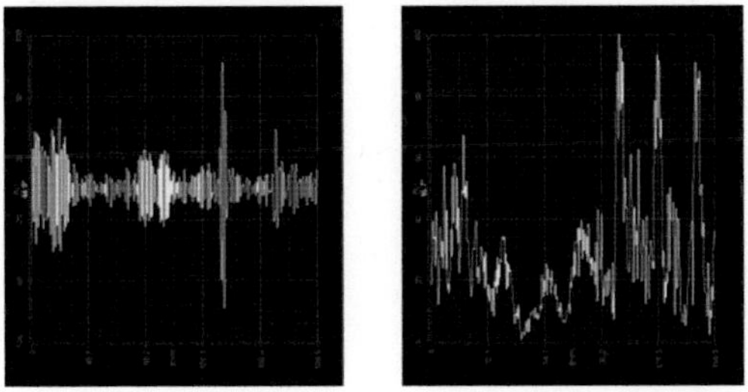

Figure 2.4: A-Scan displays: radio frequency (left) and bipolar rectified signal (right)

A B-scan represents accumulated A-scans in dimensional view. Typically, the scan position and sound wave path are displayed on the horizontal and vertical axes, respectively. The image is typically produced by a trigger gate on the A-scan. Once the signal intensity is sufficient to trigger the gate, a point is produced on the B-scan. The gate is triggered by the signal reflecting from the reflector within the material. The number of A-scans in a B-scan view is given by [24]

$$\text{Number of A} - \text{Scans} = \frac{\text{Scanning length (mm)}}{\text{Scanning resolution (mm)}} \tag{2.7}$$

In automatic ultrasonic testing case, for imaging of B-scan it is required to convert the amplitude changes of A-scans and then to make a mapping.

Considering the incident angle and time-of-flight value and the amplitudes, the mapping based both on beam locus and on amplitude can be performed and it is called "UT amplitude/beam locus mapping" (BLM) [25]. BLM is used for a basic algorithm to construct B-scan image as shown in Figure 2.5.

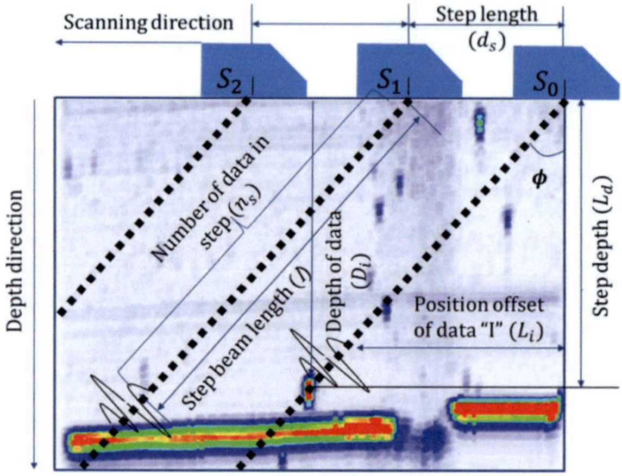

Figure 2.5: A schematic algorithm image of UT amplitude/beam locus mapping (BLM)

The number of data points acquired, n_s, can be determined as below:

$$n_s = \frac{d_s f_s}{\sin\theta c} \tag{2.8}$$

where d_s is the scanning step distance, θ is the angle of incidence, f_s is the sampling frequency, c is the sound velocity in the media.

Assuming that a certain point in the sequencing scanning lines S_x $(x = 0, 1, 2, \ldots)$ is i, the relevant scanning location L_i and depth direction D_i of the data point can be determined as below:

$$L_i = l_s x + \frac{i}{f_s} c \cdot \sin\theta \tag{2.9}$$

$$D_i = \frac{i}{f_s} c \cdot \cos\theta \tag{2.10}$$

where l_s is the step length in scanning direction.

A C-scan represents a plan-type view with the size of the specimen features. The image plane is parallel to the scan pattern. C-scan image is produced with an automated data acquisition system, for example a computer controlled scanning system. Typically, a gate for data collection is applied for the A-scan amplitude or the time-of-flight of the signal is stored as the transducer is scanned over the specimen. The relevant signal amplitudes or the time-of-flight is displayed in various color scales. The C-scan image provides information on the characteristics which reflect and scatter the sound within and on the surfaces of the specimen. Figure 2.6 shows the combined images with A-scan, B-scan, and C-scan.

A D-scan is also 2-dimensional image view and looks similar to B-scan. D-scan represents, however, the display view at right angles to the B-scan. In addition to that B-scan displays time axis versus scan axis, on the other hand, D-scan shows time axis versus index axis. The Figure 2.7 shows comparative displays of B- and D-scan which is gained from the same data acquisition.

As shown in Figure 2.8, S-scan (sectorial scan) represents 2 dimensional views which contain all the A-scans according to a time delay calculation with steered angle. The shape of the view of the S-scan is not only a pie-shape but also displays different shapes depending on software options, the properties of the window, setup condition for focal law. At a conventional pie-shape S-scan, the horizontal axis represents the width of the specimen and the vertical axis shows the beam propagation depth.

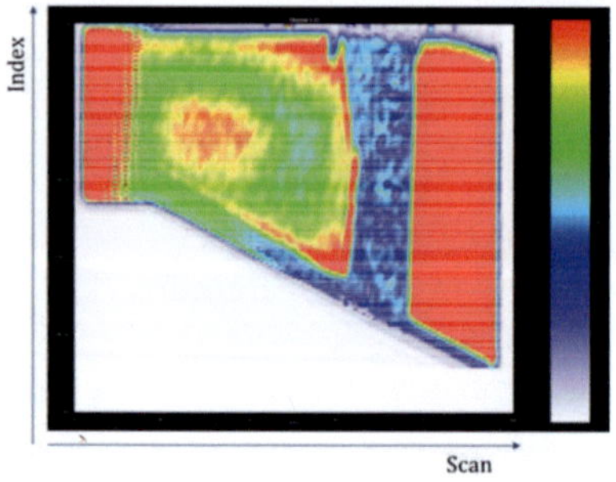

Figure 2.6: A typical image of C-scan (PCUS Pro® Array, Fraunhofer IZFP-D)

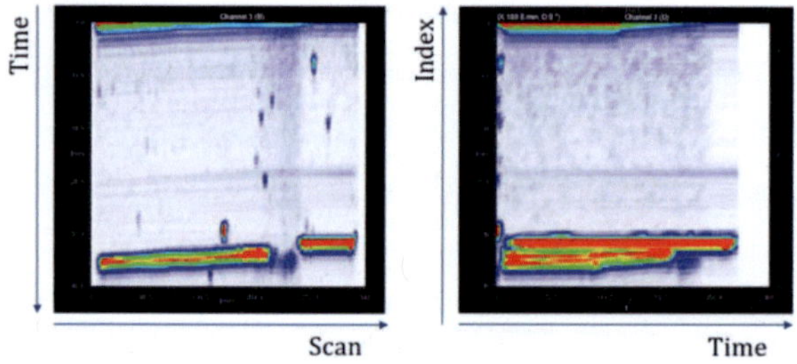

Figure 2.7: A typical image of B-scan (left) and D-scan (right) (PCUS Pro® Array, Fraunhofer IZFP-D)

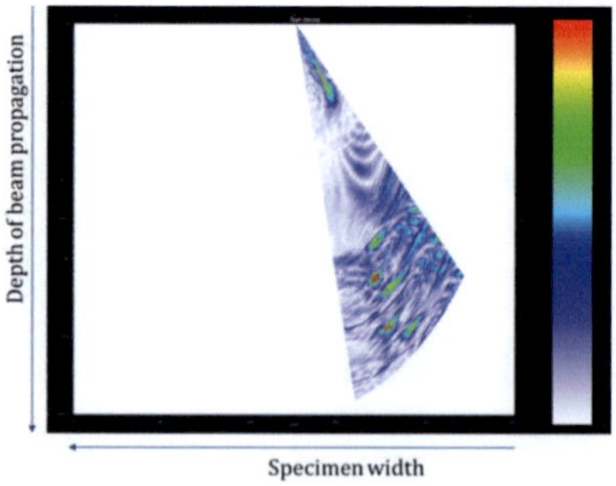

Figure 2.8: A typical image of S-scan (PCUS Pro® Array, Fraunhofer IZFP-D)

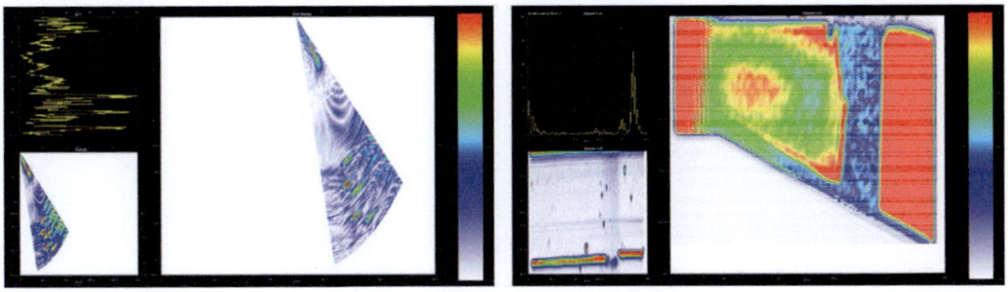

Figure 2.9: A typical multi-display image, combination of A-, B- and S-scan (left) and A-, B- and C-scan (PCUS Pro® Array, Fraunhofer IZFP-D)

Typically, multiple image views on the screen are provided at a phased array ultrasonic testing and this function offers easier evaluation of the signals acquired. Figure 2.9 shows a combined multiple image view with a spectral scale.

2.2.6 Wave distortion

When a transducer moves onto the surface of the test piece, a wave distortion can be occurred. Figure 2.10 illustrates the image distortion due to the beam width [26].

In Figure 2.10 (a), when the transducer is positioned at x_1 while the defect is in the middle of the beam width of position x_2, as shown in (b) the B-scan image will result in distorted signal mainly due to the beam width, e.g. the aperture size of the transducer, along with the different wave path in depth. To sum up, depending on how the transducer centrally locates to the reflector, the pattern of A-scans changes, i.e. the amplitude along with the time.

Such a beam distortion can be compensated by using image reconstruction techniques, for example, synthetic aperture focusing technique

(SAFT) [26] or total focusing method (TFM) [27].

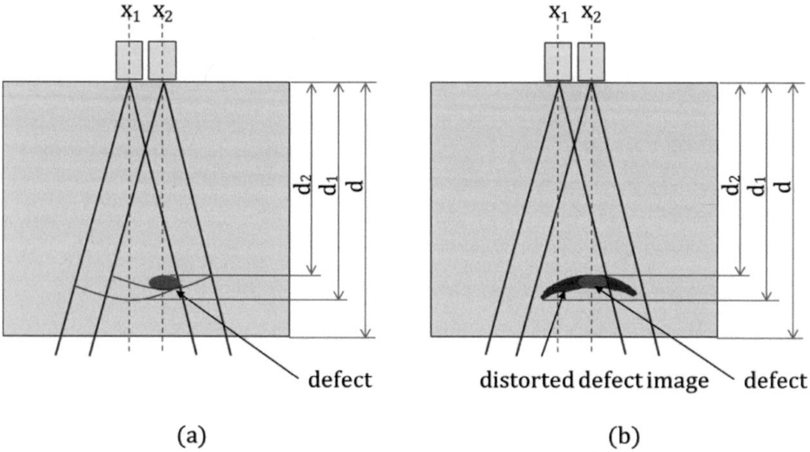

Figure 2.10: Schematic of beam distortion for (a) the defect location between transducer position x1 and x2 on the specimen (b) beam distortion image in B-scan view due to different beam width between transducer position x1 and x2

2.3 Phased array ultrasonic testing

Recently, phased array ultrasonic testing methods have been introduced in the NDT area. Main advantages of using array based testing are wider coverage of scanning range compared to a single element transducer and sophisticated evaluation of data due to electronically controlled steering and focusing function [18, 19]. In addition to that, it is also possible to make three dimensional images using two dimensional array transducers.

2.3.1 Instrumentation of phased array ultrasonic testing

All phased array instruments require the ability to configure a defined single or group of element(s) with a focal law. When the focal laws with the different values of pulser and receiver across the aperture are sequenced, the instrument creates images. A conventional diagram for a typical phased array system is shown in Figure 2.11.

While the instrument is pulsing, a trigger is fired to a bank of pulsers by sequencing time delay for a desired beam formation. At reception, the signals are then digitized and delayed by calculated the focal law. The signals are then summed to form a single RF response. Finally this waveform is amplified, filtered if required, digitized, processed and stored. As a single sequence of focal laws is being processed, at the same time the image is displayed along with associated A-Scans and its measurements.

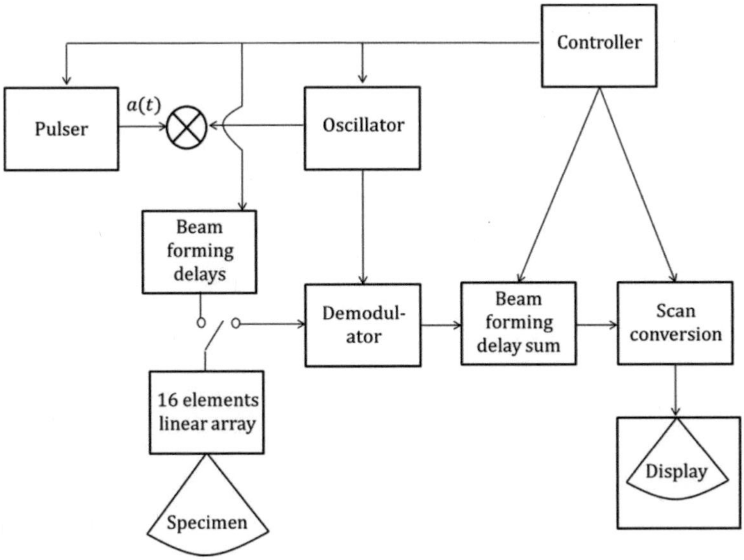

Figure 2.11: Schematic diagram of typical phased array system

2.3.2 Linear array transducers

The elements in a phased linear array transducer consists of many elements, typically 16 to 256, that are fired by signals phased for steering and focusing beam electronically [28]. The elements scan a beam electronically in the azimuth (xz) plane with the elevation (yz) plane. The nominal beam axis is the z axis. A conventional layout of the array element dimensions and the notation of the angle are seen in Figure 2.12.

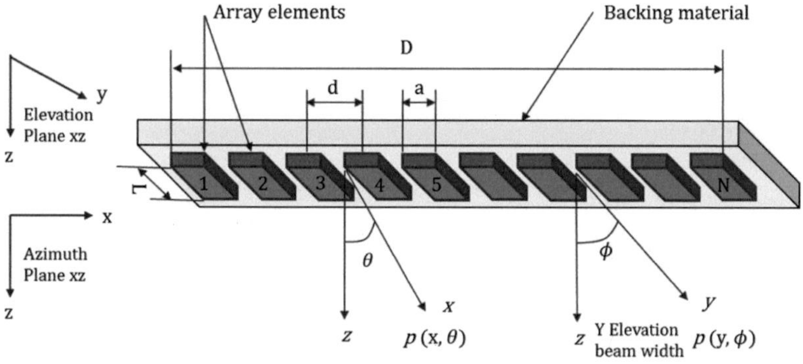

Figure 2.12: Geometry of a typical linear phased array transducer: x is scanning direction, y is elevation (L), z is nominal beam axis, p (x, θ) and , p (y, ϕ) is directivity function of azimuth plane and elevation plane respectively

The factors influencing the design of a linear array transducer are frequency (f), width of single element (a), pitch (d), the number of elements (N), and active aperture (D) which is the whole length of the array transducer and given by the following equation:

$$D = (N - 1)d\theta \qquad (2.11)$$

The thickness of each element is determined by the desired frequency. There is a rule of design for phased array transducers that the width of single element should be $\lambda/2$ to avoid grating lobes [29] at large steering angles.

In order to get an effective focusing at the maximum angle, as the beam is steered away from the plane of the array, the ability to focus the beam is reduced. The effective active aperture (D_{eff}) which can be seen along with refracted sound wave in the elastic medium is an indicator of the problem in that as the beam is steered off axis and obtained by equation (2.12) where α_I is the incident angle in medium 1 and β_r refracted angle in medium 2 [30]. The smaller effective aperture results in a shorter effective near field. The effective aperture depends on a programmed refracted angle and has a key influence on the actual focal depth and beam steering power [24].

$$D_{eff} = \frac{D \cos \beta_r}{\cos \alpha_I} \tag{2.12}$$

Minimum active aperture (D_{min}) has the aperture to get an effective focusing at the maximum refracted angle and given by equation (2.13) where c_I is velocity in first medium, c_R is velocity in specimen, f is frequency, F is focal depth for maximum refracted angle [24].

$$D_{min} = 2\left[\frac{F(c_r^2 - c_I^2 \cdot \sin^2 \beta_r)}{f \cdot c_r \cdot \cos^2 \beta_r}\right]^{1/2} \tag{2.13}$$

2.3.3 Transmitting and receiving mode

In order to produce a coherent wave with constructive interference, the transducer elements are pulsed at desired different times as shown in Figure 2.13(a).

The echo from the targeted focal point strikes the array elements with a software-controlled time shift. The echo signals arrive at individual elements and the arrival time is shifted before summation. The summed result is a form of A-scan that puts emphasis on the response from the targeted focal point and decreases the signals from other points in the material being tested as shown in Figure 2.13(b). During transmission process, the acquisition unit sends a trigger signal to the phased array unit. The latter signal is converted into a high voltage pulse accompanying a preprogrammed width and time delay calculated in the focal laws. Each element receives only one pulse which creates an ultrasound beam with a specific angle focused at a specific depth. The sound wave hits the defect and reflects back to the element.

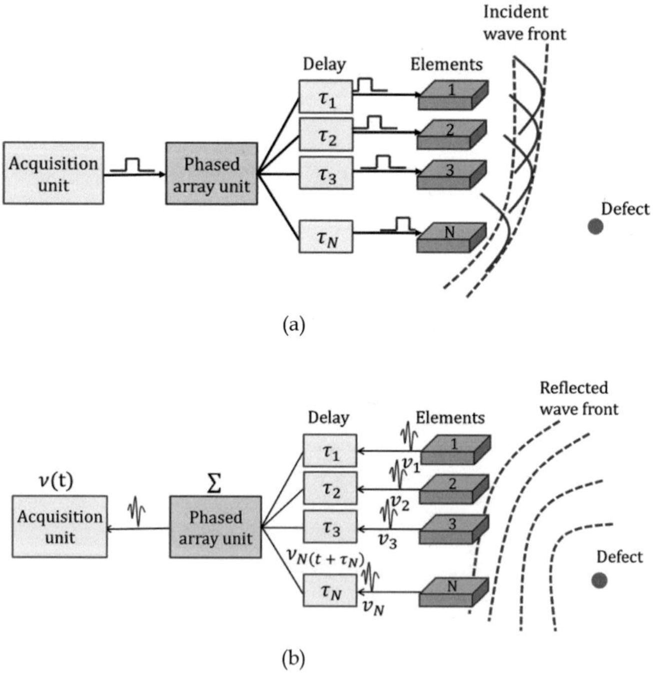

Figure 2.13: Diagram of typical array system of (a) transmitting mode and (b) receiving mode

2.3.4 Time delay setting for beam steering

In the phased array system, the time delay setting is essential for electronic scanning. In the case of the linear array transducer, the coordinate system is defined as shown in Figure 2.14.

The time delay laws relevant for each array element positioned at $x = x_n$ are expressed by the different time of the wave propagation as below [31]:

$$\Delta\tau_n(x; \mathrm{F}) = (|\mathrm{F} - x_0| - |\mathrm{F} - x|)/\mathrm{c} \tag{2.14}$$

where x_0 is the center of the linear arrays, F is focal point and c is the sound velocity. The focal point F is expressed by the incident angle θ_I. Hence, the delay law for the linear array is determined as below:

$$\Delta\tau_n(x; \mathrm{F}) = d\left[1 - \sqrt{(\sin\theta_\mathrm{I} - x_n/d)^2 + \cos^2\theta_\mathrm{I}}\right] \tag{2.15}$$

The focal distance d as in the finite limit, Equations (2.15) is simplified as below:

$$\Delta\tau_n(x; \theta_\mathrm{I}) = x_n \sin\theta_\mathrm{I} /\mathrm{c} \tag{2.16}$$

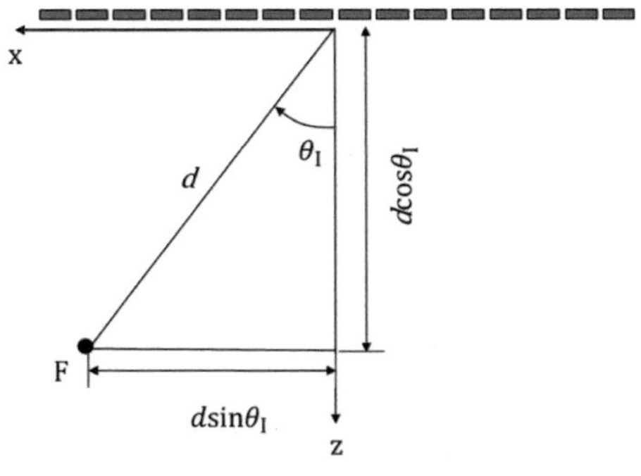

Figure 2.14: Illustration of coordinate system of linear array transducer

2.3.5 Grating lobes in phased arrays

Phased arrays produce 'grating lobes' caused by acoustic energy which spreads out from the transducer at angles other than main path. This is not only limited to phased array but also occur with conventional transducers as its size increases [19]. The first grating lobe appears at the location of $(N-1)^{th}$ lobe originated from main lobe where N the number of element.

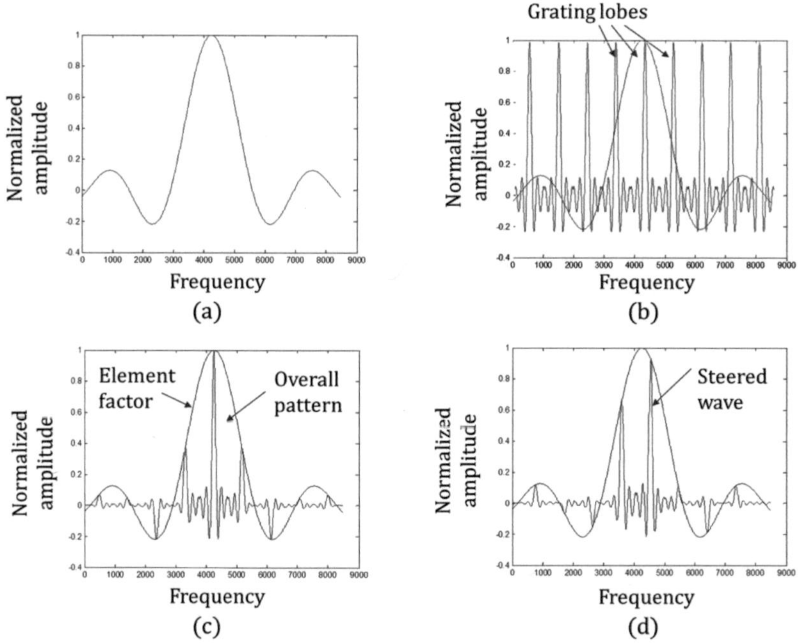

Figure 2.15: Grating lobe effects on steering describing (a) Far-field element directivity as a function of frequency for the whole element width in x axis, (b) Factors contributing to overall transform with grating lobes, (c) Modulation of grating lobes by broad angular directivity of element factor and (d) Array angular response at steered condition

The size of pitch and the elements number mainly influence the amplitude of grating lobes. In addition, in sector or angular scanning, the grating lobe is inevitably related with the steering angle. The larger the steering angle, the bigger the grating lobe(s) and Figure 2.15 [32] shows the effects of the directivity of the element on the steered wave and on the grating lobe(s).

To prevent grating lobes, the element pitch d must satisfy the condition:

$$\sin\theta = \sin\theta_s - \frac{\lambda}{d} \tag{2.17}$$

at $\theta = -90°$ is the critical element pitch d_{cr}. Therefore,

$$d_{cr} = \frac{\lambda}{1+\sin\theta_s} \tag{2.18}$$

where θ_s is the maximum desired steering angle.

To sum up, the beam steering performance can be enhanced by determining a proper inter-element spacing higher than $\lambda/2$ which is a conventional limitation [33].

Chapter 3. Specimen

3.1 Introduction

DMWs are well known not only as inhomogeneous but also as anisotropic media with highly attenuated grain characteristics. In addition, some parts of the DMWs in nuclear power plant have complicated shape and considerable thickness which require several steps of ultrasonic testing procedures resulting in increasing expense and time. Taking these obstacles into consideration, in this dissertation as a testing specimen, an outlet nozzle of the steam generator (Westinghouse Type) with the outside diameter of 31 inches (787.4mm) which is most difficult to locate cracks behind the weld was motivated. Figure 3.1 shows the image of the real outlet nozzle in the nuclear power plant.

Since it is not allowed for unauthorized personnel to take the measurements of the real component in the plant mainly due to the safety regulation of the contamination of the radioactive substance, a testing mockup was inevitably employed for this study. The testing mockup which normally implants artificial notches to verify the probability of detection (POD) of defects is basically used for a performance demonstrator for qualification of the ultrasonic testing engineer in the nuclear power plant [34]. For this dissertation, two pieces of dissimilar metal weld were cut out from the mockup by the courtesy of Korea Hydro and Nuclear Power R&D Institute. One piece of DMW mockup was used for characterization of the weld, e.g. chemical metallography and EBSD (electron back scattered diffraction) and the other piece which is named as the reference block in the study was used for visualization by SLV and for phased array ultrasonic testing. Figure 3.2 shows the mockup and the two pieces of specimen cutout from the mockup.

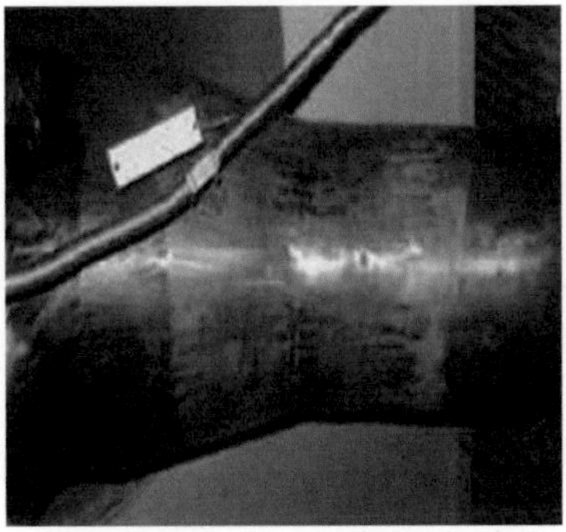

Figure 3.1: Outlet nozzle of steam generator in the nuclear power plant (Westinghouse Type, by the courtesy of Korea Hydro and Nuclear Power R&D Institute)

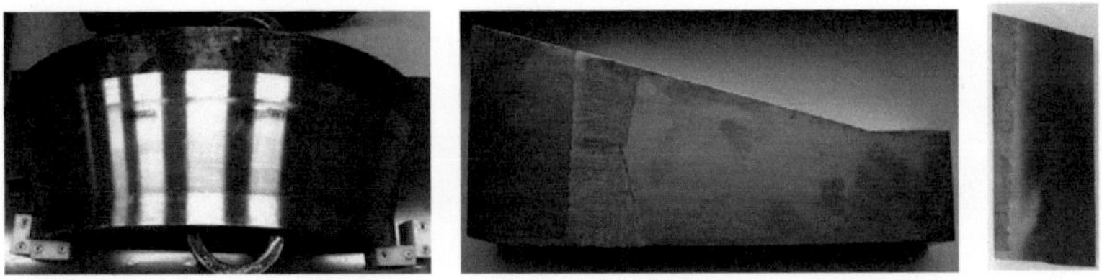

Figure 3.2: Image of performance demonstrator (PD) mockup (left), reference block (middle) and specimen for characterization of weld (right)

3.2 Profile of specimen

3.2.1 Geometrical information

Figure 3.3 shows the geometry of the specimen. As indicated, the shape is noticeably complicated which causes higher attenuation to be occurred as the transducer moves from the vertex of the cone in the pipe-safe end part toward the weld area along with the inclined surface.

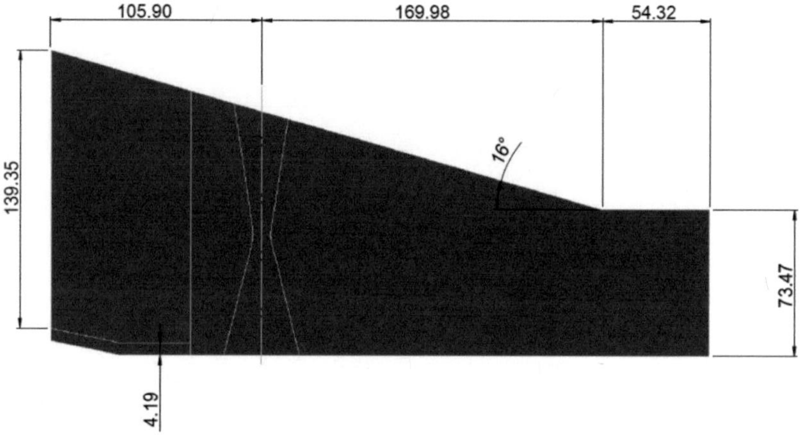

Figure 3.3: Dimension of specimen [unit: mm]

3.2.2 Compositional and weld information

The specimen has a double U-groove chamfer filled by filler material made of ERNiCrFe-7 (Inconel 600) and shielded metal arc weld (SMAW) method was applied. The weld embodies a buttering part in between the welds (ERNiCrFe-7) and mild steel (SA516). For the buttering process, gas tungsten arc weld (GTAW) method was applied and is composed of the same material as the filler material for the weld part. The Inconel filler metal produces corrosion-resistant overlays on most stainless steels. It can also be used in applications requiring resistance to oxidizing acids [35]. Figure 3.4 shows the compositional information and the content of the detailed chemical composition of the filling metal (AISI 316L steel) is shown in Appendix I.

3.3 Structural analysis methods of specimen

In this section the structure of the weld was at first analyzed by means of a scanning laser vibrometry for visualization of sound propagation in the media and for grain structures in macro scale. A metallography by using chemical etching method was performed. Lastly, electron backscatter diffraction method was fulfilled for microstructural analysis of the weld.

3.3.1 Scanning laser vibrometry combined with ultrasonic transmitting technique

3.3.1.1 Introduction

SLV is optical equipment used mainly for identifying velocity and displacement at a certain point. The laser beam is directed at the surface of the test piece on which surface should be typically mirror machined by mechanical polishing for measuring precise velocity values so as to obtain an amplitude and frequency extracted from the doppler shift of the laser beam frequency [36]. In this experiment, for obtaining beam propagation profile and grain structures, ultrasonic transmitting technique as a vibrating source of longitudinal wave is implemented with the SLV. This method has already been demonstrated for NDT applications by Schubert et al. and Barth et al [37, 38].

When the pulsed longitudinal waves start to stimulate the grains in the material, the SLV system starts to scan the surface point by point. The complete data set obtained also contains the sound wave propagation which provides the data of the grain structure extracted by the data processing [39]. Figure 3.5 shows the measurement setup of the SLV system for the visualization of sound waves in the weld using 45 degree longitudinal transducer with 2MHz.

Figure 3.6 shows the schematic diagram for the measurement. While the ultrasound longitudinal waves propagate into the specimen aiming at the target point in the weld area, the laser beam is focused onto the target point and the reflected beam is merged with reference beam and then directed onto the photo-detector. And then the interferometer compares frequency f_{mod} of the object beam with those of the internal reference beam (f_0). Lastly, the measurement of c(t) at the scanned point is subsequently proceeded by fast fourier transform (FFT).

Figure 3.4: Steel grade information on each major region in specimen

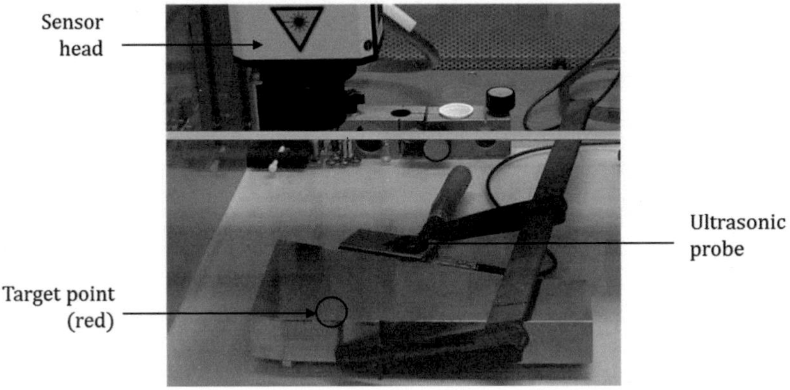

Figure 3.5: An image of measurement setup for wave visualization using scanning laser vibormetry (SLV) with ultrasonic transmitting technique

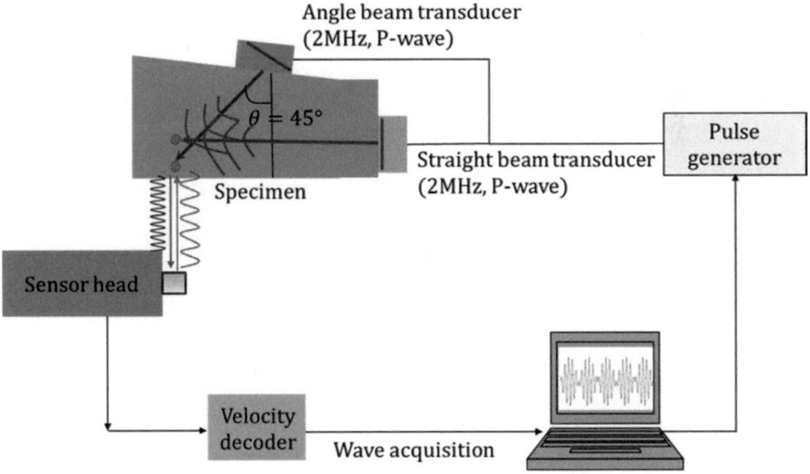

Figure 3.6: Schematic diagram of experimental setup for scanning laser vibrometry (SLV)

3.3.1.2 Visualization of wave propagation

The visualization of the wave propagation can be provided by SLV as shown in Figure 3.7. The measurement was accomplished using 45 degree P-wave transducer with 2 MHz. From the past research activities it has been reported that the longitudinal wave is normally better suited to test coarse grained anisotropic media than the shear wave at the same testing frequency of the transducer [5, 11]. In addition, because mainly of anisotropic nature, the angular difference between the group velocity and the phase velocity are larger for the vertically polarized shear wave than for the longitudinal wave or the horizontally polarized shear wave [40]. Taking this into consideration, a longitudinal mode was selected. When a conventional transducer and a pulse generator are used, the displacement amplitudes in out-of-plane reach less than tenth of a nanometer. Despite of an optimized laser reflection, noise is inevitably generated which reduces the signal-to-noise ratio (SNR). Averaging by repeating the pulses and calculating the average waveform, however, allows to improve the SNR.

The factor of scanning laser vibrometer is 50 nm/V. The signal amplitude is 1 mVpp (peak-to-peak) which means a particle displacement of about 50 pm and sensor noise is 30 mVpp. The intrinsic SNR was improved up to 45 dB by averaging more than 30,000 (2^{15} single measurements) per point.

Four slides of time snapshots gained by the measurements are shown in the Figure 3.7. To help understand the wave propagation in the weld, the snapshot images are translucently covered on the GIUM image with sufficient contrast. The original snapshot images are also attached under the images.

When the P-wave front passes through the boundary between the transversely isotropic and the weld region, the wave scattering and absorption are remarkably intensified and it becomes deformed as shown in Figure (a).

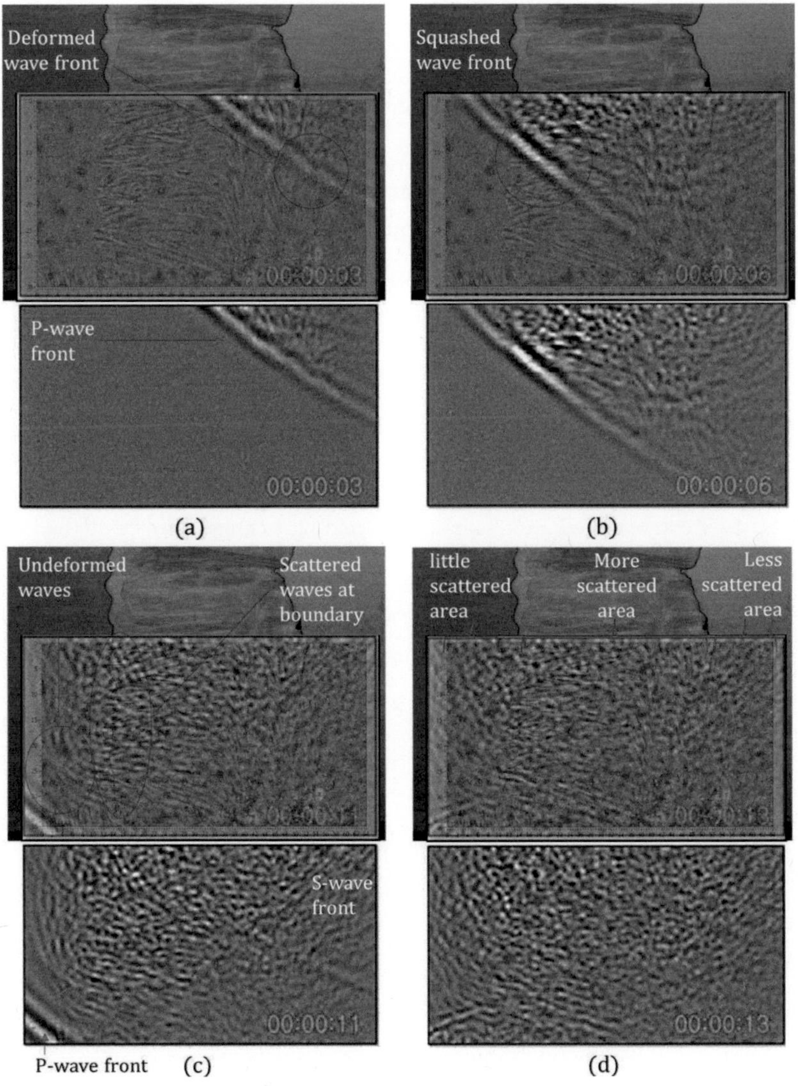

Figure 3.7: Time snapshots of the propagating waves passing through the lower part of the welds. (The elapsed propagation time of (a), (b), (c), and (d) is 3µs, 6 µs, 11 µs and 13 µs, respectively.)

When the P-wave keeps propagating into the buttering region, the wave front becomes squashed until it passes out of the region as shown in Figure (b). Once the P-wave passes through the ferrite material region, it becomes back to undeformed wave as shown in Figure (c). In the (c) image the weak S-wave propagating is found. The strong scattering signals are found in the overall weld area as shown in Figure (d). In the transversely isotropic region, on the other hand, relatively weak scattering occurs and in the ferrite material region little scattering is found.

3.3.1.3 Grain structures obtained by grazing incidence ultrasound microscopy (GIUM)

SLV combined with ultrasound transmitting technique can be applied for obtaining grain structures. To achieve this, grazing incidence ultrasound microscopy (GIUM) has been introduced by M. Barth et al. [41]. The basic idea of the technique is to obtain the grain information caused by the disturbances of the waves.

It is well known that the sound velocity and acoustic impedance represent the properties corresponding to the atomic lattice orientation. When the material is anisotropic and/or inhomogeneous, the variations in transverse expansion and the sound velocity which render the wave into out-of-plane amplitude and delay occur. In addition, when the longitudinal waves pass through grain boundaries, the waves become scattered which also facilitate to induce the out-of-plane displacement. As a result, the disturbances can be converted into the grain information in a greyscale by using a filter algorithm. Figure 3.8 shows the result of the GIUM image gained from the lower part of the weld in the specimen.

The GIUM image is translucently overlaid on the photo of the specimen representing sufficiently correlated interfaces both between the transversely isotropic and the weld regions and the buttering and the ferrite regions as marked in the blank boxes in red. The image shows the grain orientations in major pattern clearly of which the grain structures represent strong anisotropy and heterogeneous pattern. The black dots distributed over the whole area are artifacts caused by noise signal which is originated from the center of the measurement point. The dots appear in the filtering and feature extraction process when the time of wave propagation could not be found in noise signals.

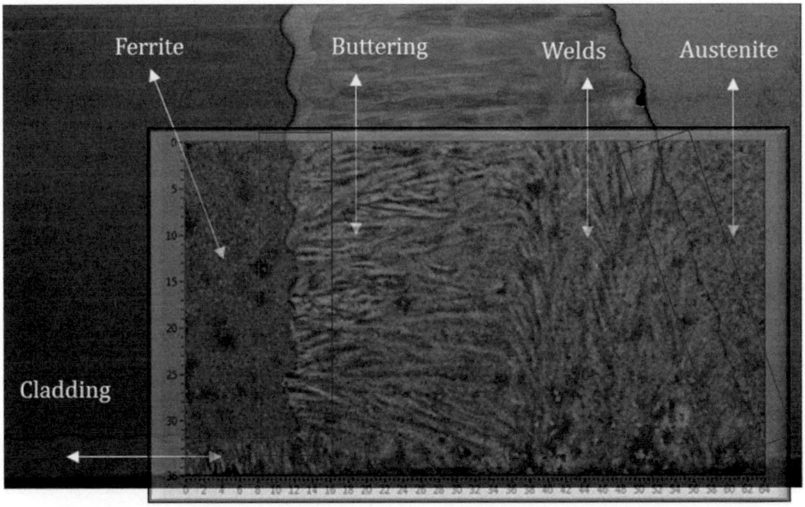

Figure 3.8: An image of grazing incidence ultrasound microscopy (GIUM) measured from lower part of weld of specimen representing major grain orientations and noticeable interfaces both between transversely isotropic and weld region and between buttering and ferrite regions

3.3.2 Chemical etching metallography

In addition to the GIUM technique, a typical macro-etching for the dissimilar metal welds was applied. One piece of the dissimilar metal weld cutout from the mockup was prepared for chemical metallography as can be seen in Figure 3.9.

The macro scale photos allow to identify patterns of the major grain orientation in dendritic foam in the weld region and in horizontal branch pattern in the buttering area. In order to investigate more precise analysis of the grain structure in the welds, electron backscatter diffraction (EBSD) technique was accomplished and described in the following section.

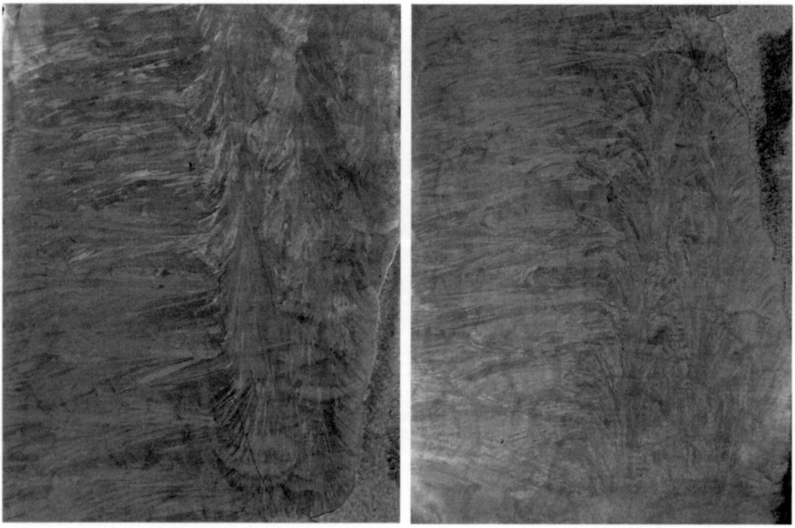

(a) (b)

Figure 3.9: Metallography achieved by chemical etching method (a) upper part and (b) lower part of the weld

3.3.3 Electron backscatter diffraction technique (EBSD)

3.3.3.1 Introduction

EBSD is a scanning electron microscope (SEM) based technique for the measurements of crystal orientation, grain misorientation, grain size and boundary characterization, phase identification and transformations, strain and fracture analysis, global and local texture, and so on. Compared to EBSD, the macro-structural analysis obtained from an etching technique has an advantage of relatively easy preparation and a relatively large measurement area. However, it has also disadvantage of impossibilities to analyze crystal orientation and the characterization of the grain boundary [42]. EBSD can be a powerful tool to provide quantitative data for modeling a complicated weld structure. Figure 3.10 shows a typical image of the EBSD measurement setup.

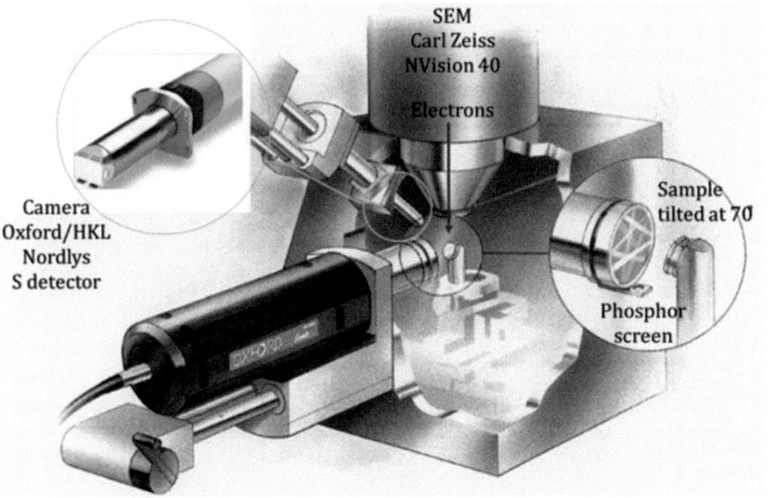

Figure 3.10: An image of electron backscatter diffraction (EBSD) measurement setup (Fraunhofer IKTS)

With the equipment used a sample can be analyzed normally within an area of 25mm2 by using SEM. The analysis of the microstructure of the fine grain under 1nm resolution is available and both the crystal orientation and the characterization of the grain boundary under the accuracy of 0.1 degree angle can be quantitatively measured. Due to the limitation of field of view, however, in order to obtain a statistically reliable data it is necessary to observe sufficient number of grains and it needs very long operation time [43]. By using EBSD, meanwhile, it is possible to analyze crystal texture through measuring relatively large area of the crystal orientation which can be observed by SEM. At the same time, it provides the quantitative analysis of the refined structure based on the change of crystal orientation [44]. Recently, the EBSD system integrating a field emission scanning electron microscope is becoming more common and that allows the analysis of the grain boundaries in nano meter scale [45, 46].

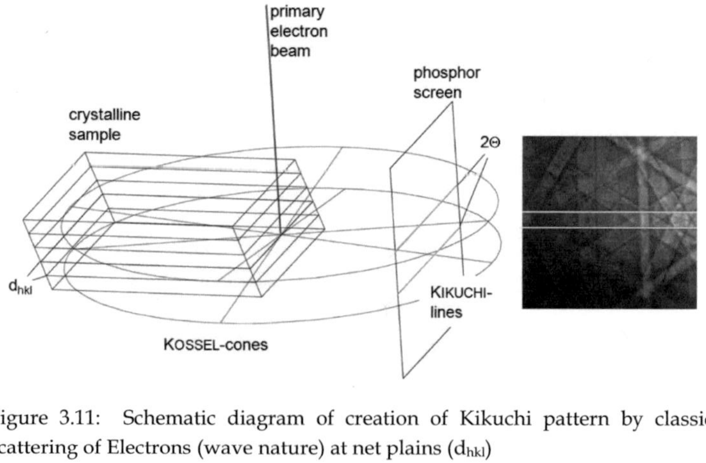

Figure 3.11: Schematic diagram of creation of Kikuchi pattern by classic scattering of Electrons (wave nature) at net plains (d_{hkl})

When the electron beam hits a sample, the diffracted electrons shape a Kikuchi pattern on the phosphor screen which follows the Bragg condition [47].

The Figure 3.11 illustrates schematically how Kikuchi pattern is created.

3.3.3.2 Preparation and measurements

The results of the EBSD analysis is mainly dependent on the surface finish condition [44]. If the deformation layer, oxidized layer, or contamination layer resulting from the preparation process of the sample is not controlled within 10 nm, it is difficult to obtain a distinct pattern and to analyze it. Therefore, in order to minimize the foreign layer different from the original sample, the deformation layer during the machining and the polishing must be as thin as possible and the surface oxidization and contamination should be minimized. Since the EBSD measurements are expensive and it takes significant time, only six pieces of sample were measured. The samples were cutout from the region where the ultrasound wave front passes through the interested interfaces, e.g. between the transversely isotropic region and the weld and between the weld and the buttering region. The region of EBSD measurement is slightly overlapped with the area of GIUM and both GIUM and EBSD results cover sufficient ROI for investigating the grain characteristics of the weld. Figure 3.12 shows the six pieces of sample chosen for EBSD measurements.

At this experiment, the samples were embedded in epoxy resin and multi-level mechanical grinding and polishing steps were performed. Finally the samples were etched with "Adler's Reagent" (Etchant). Then the sample was loaded on the SEM equipment "Carl Zeiss NVision 40". The samples then were raster scanned by the electron beam in a grid pattern with data collected with scanning resolution of 30μm. One piece of EBSD image has 4,256 points (76x56). For data acquisition "Flamenco V5.11 (Oxford)" and for data processing "Project Manager V5.11 (Oxford)", respectively was used.

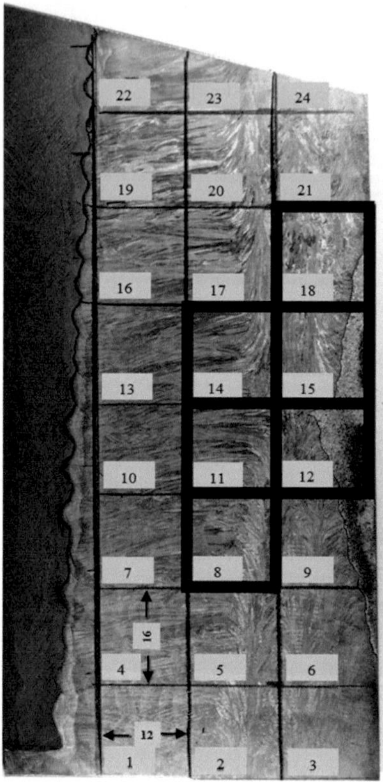

Figure 3.12: Six pieces of sample chosen for EBSD measurements (Sample No. 8, 11, 12, 14, 15, 18)

3.3.3.3 Results

The six pieces of EBSD image are shown in Figure 3.13. The weld represents a strong fibre with the vertical direction which is consistent with the observed solidification growth. In terms of size of the grain, the major grains are in the range of 2 to 4mm and are tilted from the vertical axis by up to 45°. The grains in the buttering are mostly horizontally oriented.

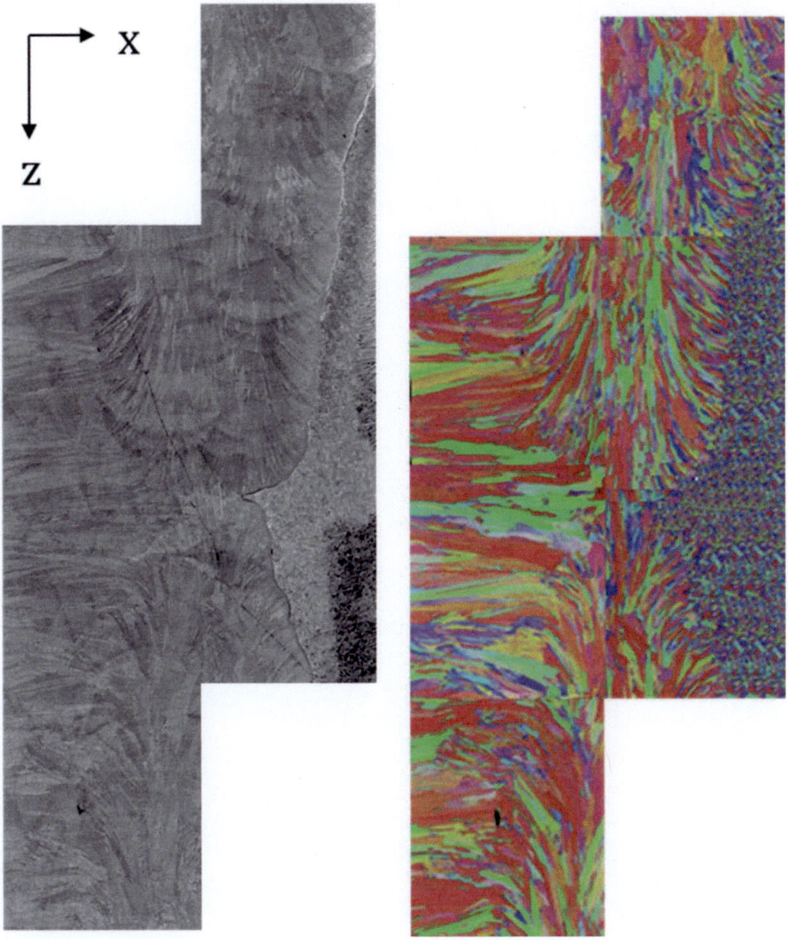

Figure 3.13: Grain structures of weld gained from six samples by metallography (on the left) and EBSD measurements (on the right)

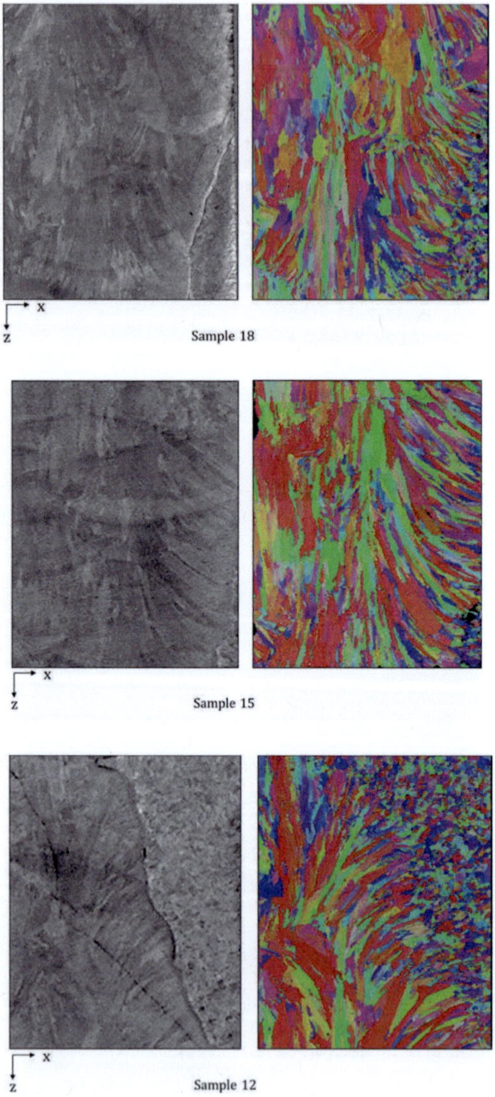

Figure 3.14: Comparison between metallography (on the left) and EBSD (on the right) for Sample No.18, No.15 and No.12

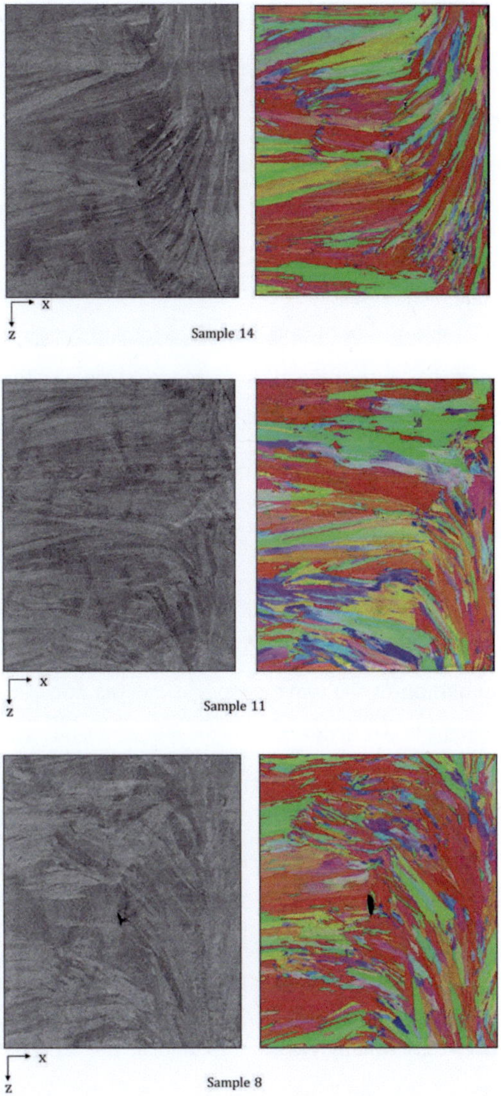

Figure 3.15: Comparison between metallography (left column)) and EBSD (right column) for Sample No.14, No.11 and No.18

3.3.4 Simulation of wave propagation in homogeneous and isotropic media

3.3.4.1 Introduction

In homogeneous isotropic media, where the time of flight is directly proportional to the wave path length, which is represented by a straight line passing through the material with a constant isotropic velocity everywhere within the media which is called Fermat's principle of least travel time [48]. In order to help understand how the wave propagates through the isotropic and homogeneous material, a numerical approach is useful. With this motivation, LS-DYNA, a commercial explicit finite element solver [49], was used in this study which has advantages of reducing computational time significantly compared to other FEM solvers.

3.3.4.2 Finite element method by using LS-DYNA

The reference block was modelled as an isotropic and homogeneous media and a simulation of the wave propagation was performed by using LS-DYNA. A 16 element linear array transducer with the same specification as the transducer used for the phased array ultrasonic testing was modelled by the FE method [50]. Figure 3.16 shows the pulse shape used for excitation of the specimen. The finite element models are discretised by the mesh size of 0.2 mm corresponding to a spatial resolution of $\lambda/20$ where λ is the wave length [51]. FE models are then applied to the mapped mesh by using 2D plane elements (four node, SECTION SHELL/Element formulation13 in LS-DYNA).

To steer the angle aiming at Notch A and B, the time delay values were determined by Eq. (2.16). The values calculated for Notch A and B are shown in Figure 3.17. The delay values and the data used for the modelling of the transducer and material are listed in the Appendix A.2.

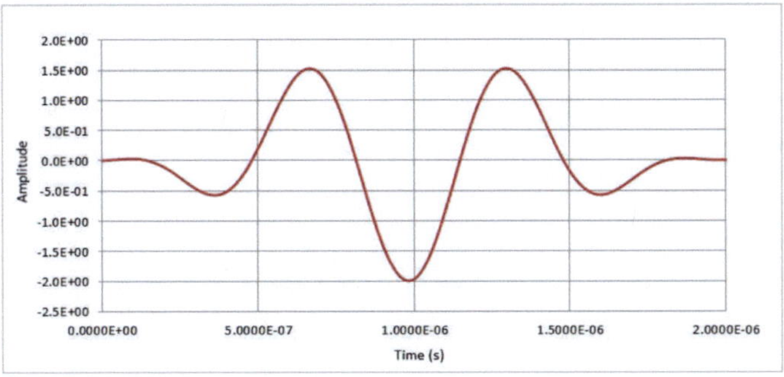

Figure 3.16: Modelling of piezoelectric transducer with equivalent force (1.5MHz)

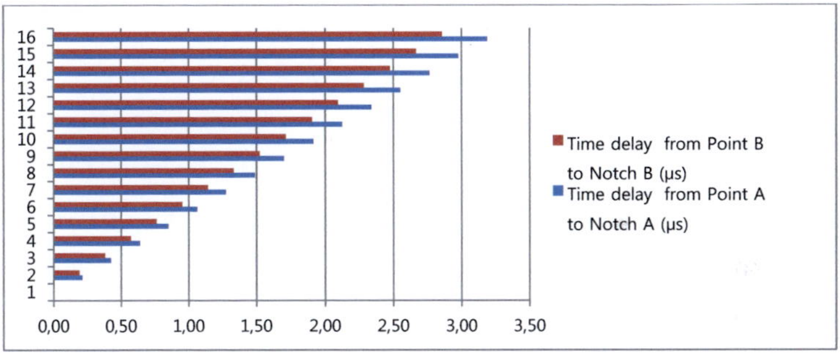

Figure 3.17: Time delay values calculated for Notch A and Notch B

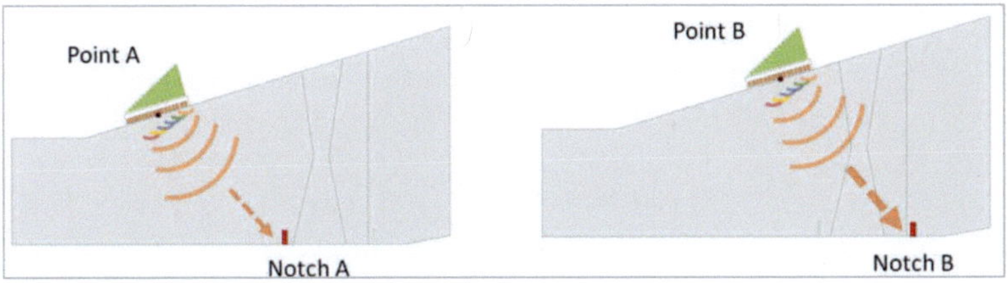

Figure 3.18: A schematic illustration of two FE models for detecting Notch A and Notch B

For scanning each of Notch A and B, each element is excited by the same transient pulse and the calculated time delays in sequence. Figure 3.18 illustrates the scanning and the focusing scheme for both notches.

3.3.4.3 Results

In the simulation, it is assumed that the ultrasonic waves are transmitted into the specimen by a linear array transducer as shown in Figure 3.19. The delay time of the transducer is set to focus the P-wave to the tip of Notch A and B. The P-waves arrive at the notch tips and then the diffracted waves from the tips are generated. It is clearly shown that the P-waves arrive at the two notch tips as calculated and the wave fronts are not deformed and skewed. In addition, little scattering was occurred.

Apart from the P-waves, shear (S) waves also travel behind the P-wave. However, the S-waves were not of interest since they were not focused to the target notch. Hence, the S-waves arrived at the bottom in front of the target notches. It is also noticed that surface waves are generated along the surface of the specimen and propagate away from the wave source. Due mainly to the finite element method by rectangular discretisation grid, the roughness of the grids was inherently generated onto the inclined surface. Such a roughness caused the surface waves to render some blur noise as marked in white circles in Figure 3.19. This noise would be reduced or eliminated by other types of grids, i.e., triangular grids.

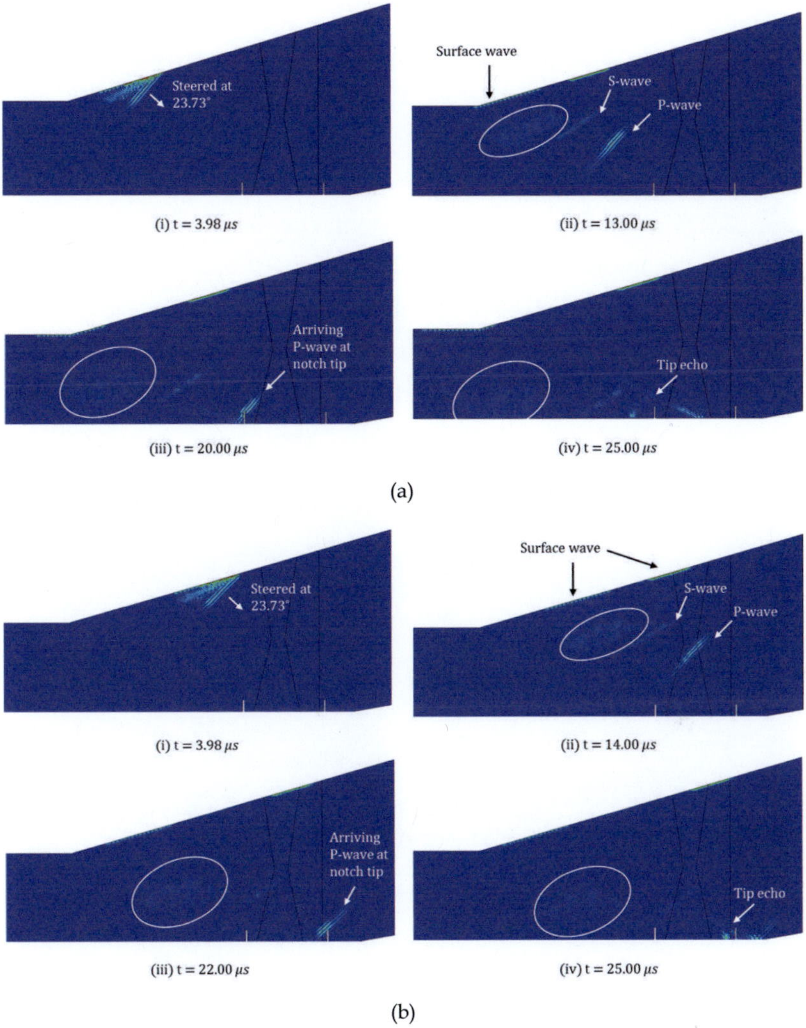

Figure 3.19: Simulation results of ultrasonic wave propagation to notch A (a) and notch B (b)

3.4 Production of artificial flaws

It has been investigated that most cracks are often created in or around heat affected zone (HAZ) of the weld root area of the inner side. The most difficult location to detect cracks, of course, is the place behind the buttering area mainly due to the high attenuation in both the welds and the buttering area. Therefore, the location of the sound reflector for demonstrating crack is determined in the place which is called as "Notch B" in this experiment. The type of the artificial flaw is electron discharge machined (EDM) notch and two identical notches are produced through the whole thickness. Figure 3.20 shows the dimension and the geometric information of the notches.

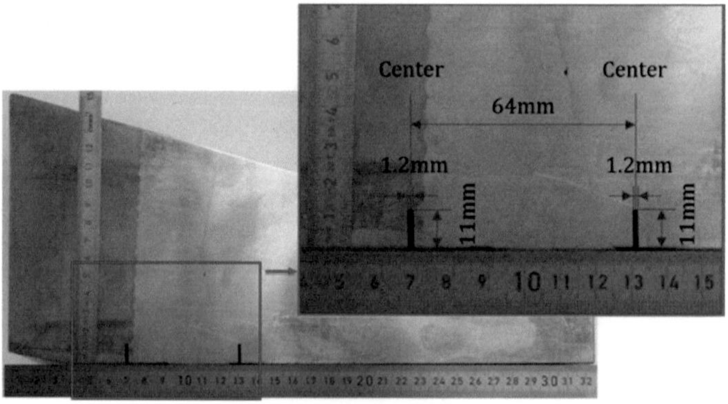

Figure 3.20: Dimension and geometric information on EDM notches

In this dissertation, the key idea is to make use of a reference data obtained from Notch A and B in the specimen. A basic assumption of the idea is that the wave propagation passing through the transversely isotropic media may render no significant variation in the beam direction since the microstructure in that media does not vary evidently as the wave runs [52, 53]. This assumption was proved by using Notch A by the referencing delay law technique in Chapter 5.

3.5 Ultrasonic immersion testing

After producing the EDM notches, in order to confirm if the specimen contains any defect other than the EDM notches, an ultrasonic immersion testing method was applied using PCUSpro Multi ultrasonic device with 2 MHz pulse-echo longitudinal transducer for immersion testing. The pulse length 125ns, gain level 6.0 dB and the sample rate was 80 MHz. In order to scan the whole volume of the specimen, the specimen was placed in the cross section as shown in Figure 3.21 and was scanned in X and Z-axes.

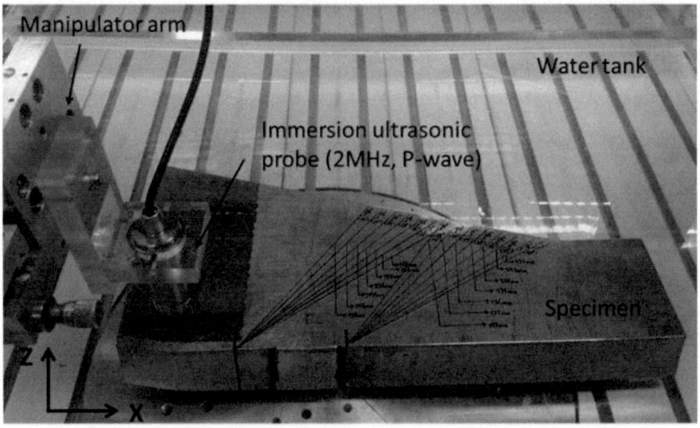

Figure 3.21: An image of the immersion tank ultrasound testing setup

After scanning the full range of the specimen, the B- and C-scan images in the spectral scale were obtained as shown in Figure 3.22.

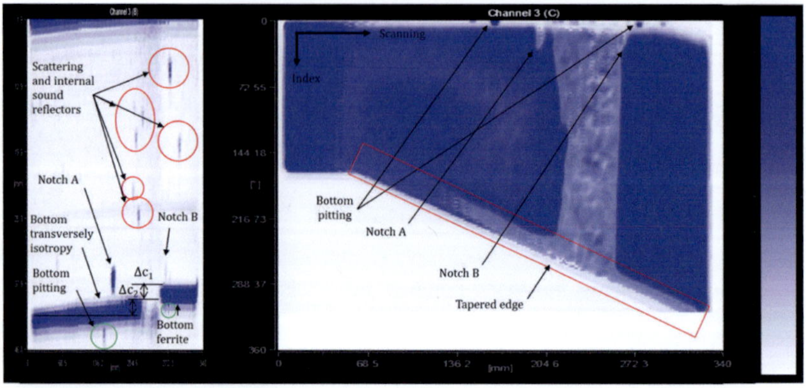

Figure 3.22: Presentation of specimen with the evidence of EDM notches, B-scan image (left) and C-scan image (right). Notch B signal is relatively week compared to notch A signal. Scattering and internal sound reflector signals are found in the weld. A noise signal due to a pitting on the bottom is detected. Due to the difference of the sound velocities of austenitic and ferritic media, the images reflected from each bottom show the difference of the time of flight.

Notch A was detected despite of some blur effect. Taking the scanning direction into consideration, due to narrow lateral distance between the weld and the Notch, Notch B was not disclosed distinctly in C-scan image but in B-scan image the signal was identified despite of weaken intensity. Since the absorption and scattering in the austenitic weld area are strong, some scattering signals were detected. Among these signals which were interpreted as scattering, one signal was probed during the phased array ultrasonic testing in Chapter 5. After investigation, the signal has been suspected as a sound reflector which would be caused by an internal inclusion-like flaw in the weld. It is interesting to notice that mainly due to the difference of the sound velocity between the transversely isotropic region and the base metal (ferrite), the signals in the two regions appeared in different time of flight (Δc_1) in B-scan image. The bottom signals in the transversely isotropic region were foamed as if the contour of the specimen is inclined in B-scan image. However, the surface is flat and the reason of the fluctuation Δc_2 would be assumed that the sound velocity in transversely isotropic media is not even both in scanning and in index directions. On the outer surface in the bottom there are weld pittings which are shaped in a ball type. These were found both in B- and C-scan images. Due to the geometrically tapered contour at the edge in the inclined part, some blur signals were occurred as shown in C-scan image.

To sum up, the total volume investigation by the immersion ultrasonic testing can provide general information on the specimen but would not be sufficient to quantify defects.

Chapter 4.
Data acquisition and image processing

4.1 Data acquisition by full matrix capture (FMC)

4.1.1 Introduction

Full matrix capture (FMC) is a specific technique to acquire signal data using phased array ultrasonic transducers which has a similar algorithm to sampling phased array [8]. This technique provides a complete time domain signal set which is received from every element of an array transducer. This technique was introduced by Holmes et al. in 2005 [54]. Since then, FMC has been mainly applied in the medical ultrasound imaging area [55]. K. Thomenius introduced FMC as the "gold standard" to which variant methods can be compared [56]. The FMC has also been used for inspection of isotropic carbon steel material [57]. In the FMC technique, one element shots while the other elements are receiving and this procedure is repeated until all elements are shot, in which FMC does not need to define the steering condition and focal laws before the testing. The output is usually a huge matrix, e.g., $n \times n$ for an array of n elements. Such a huge amount of data set can be a disadvantage on processing speed. Practically FMC needs image reconstruction methods. Nevertheless such a sufficient data allows ultrasonic testing to acquire all possible data.

The characteristics of the technique are 1) every pixel has ideal focus on transmitter and on receiver 2) beam former parameters, e.g., velocity, defined at reconstruction 3) array, interfaces and target geometry are defined when processing 4) integrated scan-conversion 5) Predefined focal laws and steering is not necessary which results in avoiding grating lobe(s). Due to these characteristics, FMC has advantages of enhancing the sensitivity for small size defects and increasing the region of interest within the range of scanning [54, 58].

The Figure 4.1 illustrates the sequencing of the shots fired by a virtual five elements phased-array transducer based on FMC (red: transmitting element, blue: receiving elements). The data are collected in a matrix S that contains all the acquired signals. S_{ij} contains the signal transmitted by element i and received by element j. The tx and rx represent transmitter and receiver elements, respectively and each subscript means the sequence of element number to be fired.

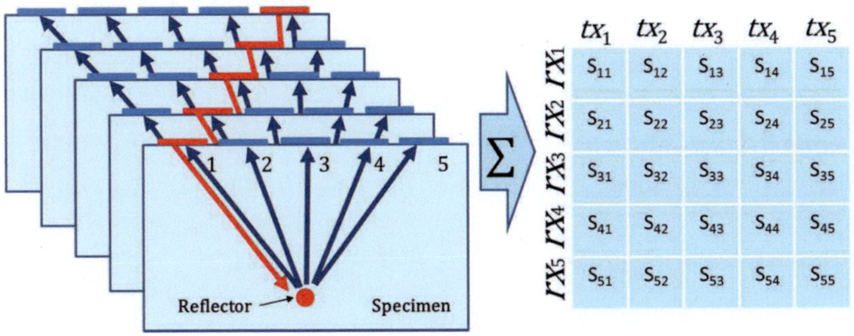

Figure 4.1: Schematic diagram of algorithm of full matrix capture

All the signals acquired from every single array element, as an input data-set, are processed for the image reconstruction using TFM algorithm discussed in 4.2. Since the sound field of a single array element is divergent, every time signal received contains overlapped echo-signals reflected from the flaw in different positions. The reconstructed image in one position of linear array visualizes as a B-scan.

4.1.2 Principle

For the data acquisition through the FMC, the axis of coordinates is shown in Figure 4.2. The distance of the wave propagation can be determined by the equation below:

$$d_{tx,rx} = \sqrt{(x_{tx} - x)^2 + z^2 + (x_{rx} - x)^2 + z^2} \qquad (4.1)$$

Each of x and z is the coordinate of the circular reflector, like a side drilled hole and each of tx and rx represents transmit and receive element, respectively.

When the phase shift is applied for the output function, f(t), a simulation for the wave propagation can be achieved. The period of beam propagation, t, can be determined from $d_{tx,rx}$. The frequency spectrum, $F(\omega)$, can be calculated from the output signal of the element by the Fourier transform and is given as the following equation:

$$F(\omega) = \int_{-\infty}^{\infty} f(t) e^{iwt} d\omega \qquad (4.2)$$

Then, the complex spectrum of the signal phase shifted, $G_{tx,rx}(\omega)$, is given as:

$$G_{tx,rx} = F(\omega)e^{(-i\omega/c_1)d_{tx,rx}} \tag{4.3}$$

In order to include the directivity of the component and beam dispersion, the spread spectrum is multiplied by amplitude factor. McNab and Stumpf defined the directivity as the equation below [59]:

$$p(\theta,\phi) = sinc\left(\frac{\pi a \sin\theta \cos\phi}{\lambda}\right) sinc\left(\frac{\pi L \sin\theta \cos\phi}{\lambda}\right) \tag{4.4}$$

Where a is the width of single element, L is the length of the element, λ is the wave length, θ is the steering angle against the vertical line of the element and the ϕ is the angle between the vertical line and elevation plane. In the normal case of phased array ultrasonic testing, the relationship between L and a is $L \gg a$ and therefore the directivity functions for both transmitting and receiving elements can be written as follows:

$$p_{tx} = sinc\left(\frac{\pi a \sin\theta_{tx}}{\lambda}\right) \text{ and } p_{rx} = sinc\left(\frac{\pi a \sin\theta_{rx}}{\lambda}\right) \tag{4.5}$$

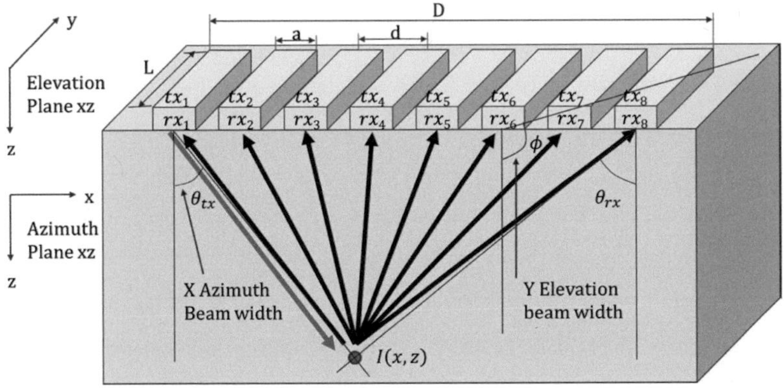

Figure 4.2: An illustration for algorithm of full matrix capture (x is scanning direction, y is elevation (L), z is beam axis.)

For modeling how the ultrasound waves fired from the transmitter and reflected from the reflector are divergent, the 2-D inverse power law can be used and the amplitude, $A_{tx,rx}$, passing through d_{tx} and d_{rx} is determined by given equation below:

$$A_{tx,rx} = \frac{A_0}{\sqrt{d_{tx}, d_{rx}}} \quad (4.6)$$

where A_0 is the signal amplitude received from a unity propagation distance.

Every pair of transmitter and receiver acquires maximum n^2 time-domain signals. When each element has the same performance over the whole array, some redundancy with the acquired data can occur. [58].

4.2 Data processing by total focusing method (TFM) combined with full matrix capture (FMC)

4.2.1 Introduction

The TFM has advantages of improving SNR and lateral resolution which does not require a traditional ultrasonic lens for focusing [60]. When the sound wave is focused to a focal point in the region of interest, the resulting image provides sufficient resolution. Such a focusing can be typically performed by three methods. The first method is so called "dynamic depth focusing" which is operated at the transmission stage with the calculated delay time of each element. In a conventional phased array ultrasonic testing system, each element is individually pulsed at calculated different times for beam focusing with desired focal depth. Such manipulation allows dynamically focusing and steering the beam at one or more desired points [61].

The second method is named "focusing and steering' which is performed at the receiving stage by applying the same delay law as at the transmitting stage. The received signals are thus delayed accordingly and summed in phase. However, due to the inevitable relationship between steering and grating lobe(s), this method has disadvantage of increasing the grating lobe(s) as discussed in Chapter 2.3.5.

The other method is an image reconstruction at post-processing stage. The focusing can be performed by synthetically shifted and sampled A-scans. For example, inverse delay law can be adopted in the receiving mode and then the signals are time shifted accordingly and summed. In this dissertation, two different data processing methods are used, one is TFM combined with FMC and the other one TFM combined with SAFT [54, 62, 63].

In general, TFM can also be divided both into synthetic aperture focusing technique (SAFT) and into common source method (CSM). However, the algorithm is somewhat different depending on the combination of the signal acquisition between transmitters and receivers [64]. Figure 4.3 shows the different algorithm in schematic diagram.

The main difference of the algorithm of SAFT from TFM is that a transmitter sends the wave and only the same element receives the signal in SAFT. CSM also varies with respect to TFM in that the algorithm is independent of the transmitting element because the transmitter is common to all data to be scanned. In addition, CSM is relatively faster data-processing method since it takes only single shot while both TFM and SAFT require the complete process according to the number of elements.

The both TFM [54] and SAFT [65, 66] have great advantage of a higher possibility of defects (POD) by increasing SNR.

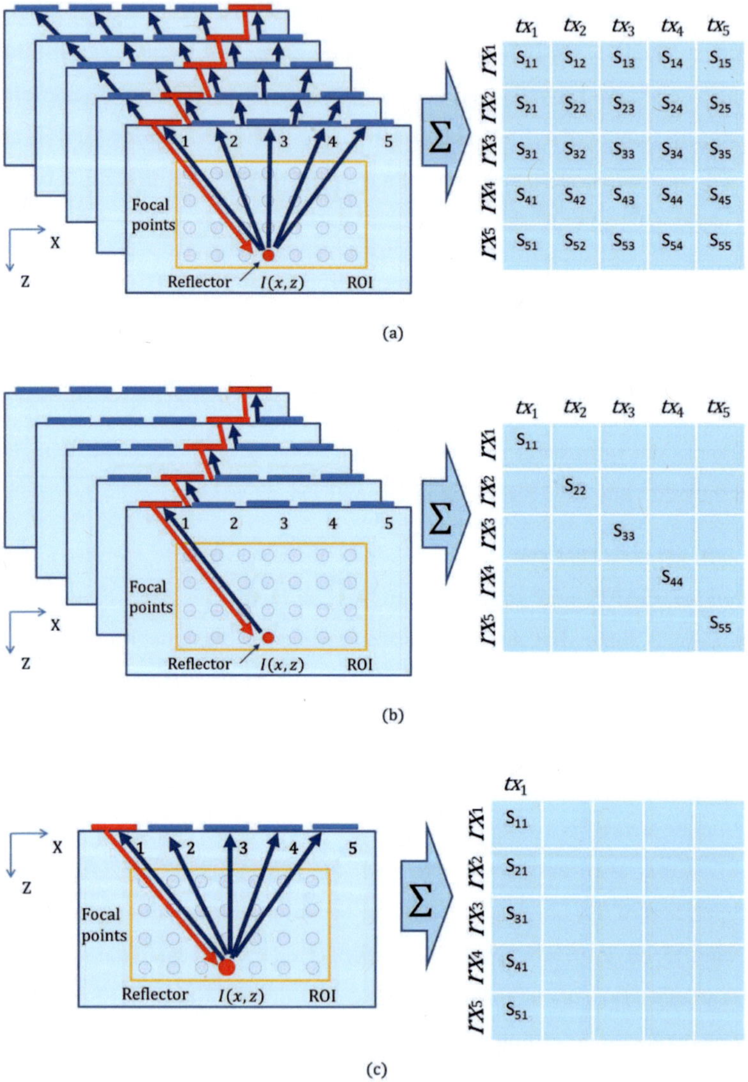

Figure 4.3: Schematic diagram of different algorithms depending on transmitting and receiving combinations (a) Total Focusing Method (b) Synthetic Aperture Focusing Testing (c) Common Source Method

4.2.2 Principle of total focusing method (TFM)

For the application of TFM combined with FMC, each transmitter is shot in sequence by FMC algorithm. The data acquired are then stored in the matrix. Once the matrix is full, all the data are ready for data processing. For the computation the region of interest (ROI) is defined for the data reconstruction. This region is then meshed and for each focal point on the grid, the focal laws can be calculated for the complete set of transmitting -receiving pair combination of the elements. All the signals are time-shifted and then summed at every point of the grid. However, typically depending on the number of focal points in ROI, selected signals are time-shifted before the summation of the desired point of the grid. The basic assumptions of this combination method for the post-processing are that the sound propagation from the transmitter to the receiver is along straight lines (subject only to refraction at the component) and that the sound velocity is constant [67].

As an example, the matrix in the Figure 4.4 represents two different focal laws calculated both for the focal point $I(x_1, z_1)$ in red and for the point $I(x_2, z_2)$ in dark green, when the focal law is determined aiming at the points. The red line in the matrix is corresponding to the target focal point in red and the dark green lines belong to the point in dark green. The time of flight values for both target points transmitted and reflected at each element were arbitrarily determined according to the geometric scale and based on the origin, e.g. the center of the first element. Using such a post-processing, a focal law value in ROI is acquired.

Taking the FMC algorithm into consideration, each element is fired in sequence and all elements receive the signal independently, in order to calculate the focal laws. The data obtained from the complete transmitting-receiving process is stored in a matrix, A_{ij}, the size of matrix is determined by "number of transmitting elements (n_{tx}) × number of receiving elements (n_{rx}) × number of samplings in each signal (n_s)". Delay laws can be different from one focal point to other focal points in the grid. The size of the matrix for the whole ROI is determined by "number of focal points in x-axis (np_x) × number of focal points in z-axis (np_z) × number of transducer elements (n_e)".

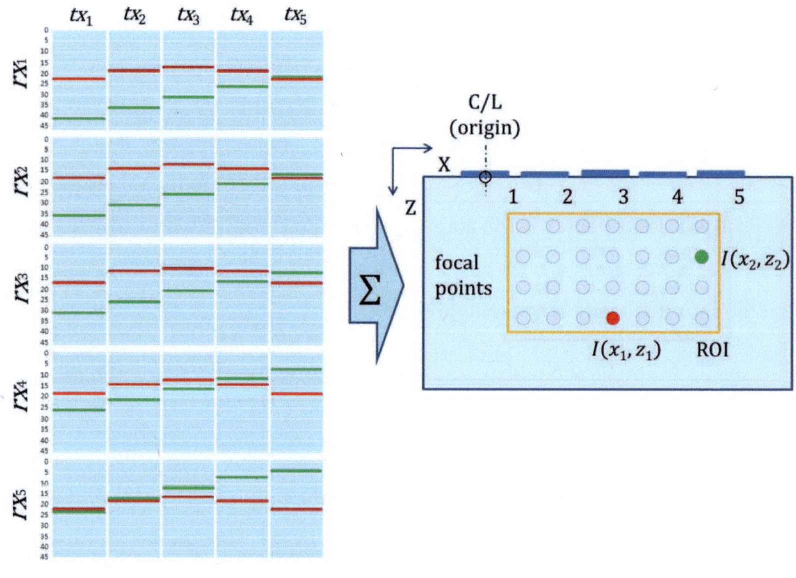

Figure 4.4: An example diagram of algorithm of total focusing method, focusing at arbitrary points $I(x_1, z_1)$ and $I(x_2, z_2)$ in region of interest

Assuming that the matrix $A_{ij}(t)$ is $1 \leq i \leq n_e$ and $1 \leq j \leq n_e$, the array is focused to produce an image matrix $I(x, z)$ as below:

$$I(x,z) = \left| \sum_{tx=1}^{n_{tx}} \sum_{rx=1}^{n_{rx}} A_{tx,rx} \left(\frac{\sqrt{(x_{tx}-x)^2+z_{tx}{}^2}+\sqrt{(x_{rx}-x)+z_{rx}{}^2}}{c} \right) \right| \qquad (4.7)$$

where x_{tx}, z_{tx} locations of transmit element,

x_{rx}, z_{rx} locations of receive element,

c is sound wave velocity in the material

4.3 Referencing delay law technique (RDLT)

4.3.1 Introduction

In ultrasonic testing, there is a need for reference blocks which is also called comparator blocks [68]. Reference blocks provide a consistency of measurements and help to interpret and to quantify the information received. The blocks should validate the reproductivity or the repeatability of the results and provide similar results produced by different ultrasonic testing systems. A reference block is normally produced by the same process as the original testing specimen and the material to be used should be similar in its sound attenuation to the material to be tested [69]. The blocks are typically manufactured with artificially induced flaws, for example side drilled holes, flat bottom holes and electro discharged machining notches. In most cases of ultrasonic testing, types of defects to be detected and the ROI are specified. The quantification of the data, for example the size of the flaws can be determined by the blocks. By comparing the signal obtained from the artificial flaw of the reference block to that received from the actual flaw of the testing specimen, estimation of the actual flaw size can be accomplished [68].

After reviewing the above mentioned, for the realistic quantitative ultrasound testing, a basic idea of utilizing such the nature of the austenitic stainless steel which is one part of the dissimilar metal weld specimen was generated. The main approach is to obtain a basic-data-set obtained from the EDM notch where the waves pass into the transversely isotropic region by FMC, and then to determine the focal laws of the EDM notch behind the austenitic weld area by referencing the data acquired. For referencing process, it is assumed that once the transversely isotropic region is corrected by the proposed technique, the signals behind the austenitic weld which are misplaced due to the beam distortion can be correspondingly corrected. This technique is named as Referencing Delay Law Technique (RDLT). The RDLT can provide realistic quantitative method through improving SNR and overcoming the degradation of the testing result due to the wave distortion. In addition, the technique can also minimize the individual difference in acquisition of testing data because it uses a full data set covering the whole ROI and focusing at any point in the grid in the area [54].

The principle with methodological description and the results which were explored for the concept using an austenitic calibration block are presented in Chapter 4.3.2. The experimental results of using the RDLT on the reference block are presented in Chapter 5.

4.3.2 Principle

As the name of the technique indicated, the key part of RDLT is to calculate directly the sound velocities in the transversely isotropic region so that the region should be referenced for correcting the misplaced signals. For this technique, it is essential to understand the characteristics of wave propagation in transversely isotropic media.

R. A. Roberts [70] reported the analytical problem of transmission of an ultrasound wave from a homogeneous, isotropic elastic half-space into a homogeneous, transversely isotropic half-space demonstrated using columnar-grained steels. In the transversely isotropic media, the material provides the advantage that all resulting energy vectors are confined to the plane of anisotropy as long as the incident energy vectors are in that plane [64]. In the study results of R. A. Roberts, the longitudinal incident wave in 0, 45 and 90 degree angle becomes distorted when the beam is transmitted into the interface of the transversely isotropic media. This would be mainly caused by the difference of the elastic properties in the grains. Elastic constants of transversely isotropic material in the austenitic steel are listed in Table 4.1 [64].

Material parameter	Value
C_{11}	249×10^9 Nm^{-2}
C_{12}	124×10^9 Nm^{-2}
C_{13}	133×10^9 Nm^{-2}
C_{33}	205×10^9 Nm^{-2}
C_{44}	125×10^9 Nm^{-2}
C_{66}	62.5×10^9 Nm^{-2}
ρ	7.85×10^3 kgm^{-3}

Table 4.1: Material properties for the transversely isotropic austenitic steel

Figure 4.5: Snapshots of ultrasonic wave propagation in a transversely isotropic homogeneous media in austenitic steel of dissimilar metal weld [71]

K. Nakahata et. al [71] modeled transversely isotropic region in the austenitic steel of the DMW. Figure 4.5 shows the snapshots of the wave fronts by means of the Elastodynamic Finite Integration Technique (EFIT) which are related to the group velocity curve. Since the modeling is set for 2D-EFIT, it is only possible to simulate the wave front under plain strain conditions.

Considering the wave behavior within a structure composed of both an anisotropic inhomogeneous and a transversely isotropic homogeneous media, e.g. dissimilar metal welds, the idea was formulated to correct the transversely isotropic region as an isotropic homogeneous region where the waves propagate along straight lines with constant wave velocity. For this, it is necessary to determine the sound velocities for at least two different directions of the wave propagation. In this study, a practical approach to approximate the velocities directly from the specimen with the aid of Notch A has been applied by using FMC-TFM technique and the method was described in Chapter 5. This process is named in this study as Referencing. Once the transversely isotropic region is corrected, the time delay values for the anisotropic inhomogeneous region, e.g., Notch B, can be calculated by referencing the region. Thus one would take RDLT for correcting the distorted region where the complex waves pass with varying velocities into a simple geometry where the waves travel in simple path with constant velocity.

To summarize, RDLT can be useful for correcting a misplaced finding in ultrasonic testing of the anisotropic media by referencing a transversely isotropic region. Figure 4.6 shows the schematic diagram of how the RDLT is processed in sequence.

At the preparation stage, a reference block is designed with relevant artificial flaws. One flaw should be located in the region where the wave propagation is relatively not affected by the attenuation and it becomes a reference flaw. On the other hand, the other flaw(s) should be placed where the defect should be detected and it is called as a target flaw. The geometric information containing the reference block and its flaws are then profiled. Finally, measurement setup including an optimum testing transducer is completed.

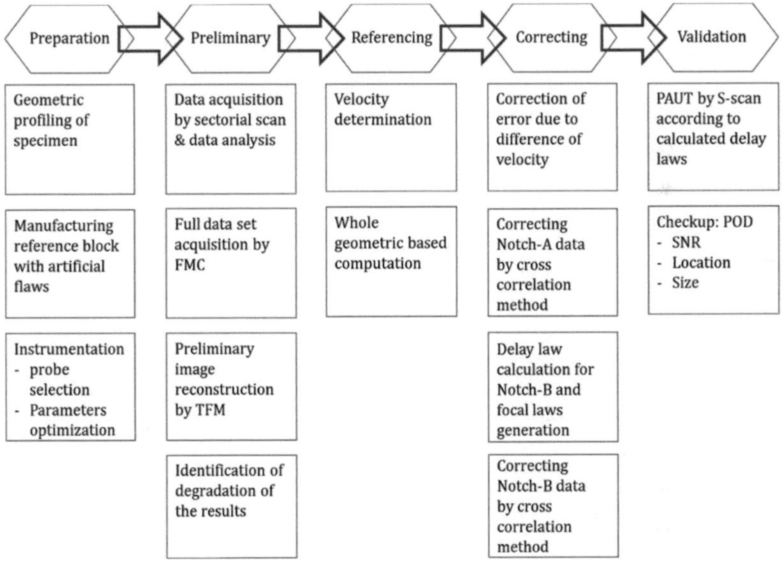

Figure 4.6: A schematic diagram of the procedure for the referencing delay law technique

In the RDLT, the reference notch is used under the assumption as if it is a typical isotropic weld and the required compensation process is accomplished at the correcting stage.

At the preliminary stage, beside the conventional phased array testing by S-scan mode, the first objective is to obtain a full data set and to reconstruct the whole volume of image in the ROI by using the FMC-TFM combination technique. Through the whole volume image gained, the degradation of the testing, for example misplacement of the flaw found, can be identified. The full data set is practically valuable not only for reconstructing image but for referencing parameters, e.g. sound velocity values which are performed at the referencing stage.

Through the referencing stage, the key parameters, e.g. the sound velocities measured directly from the specimen, are obtained. Based on such the acquired velocities and the geometric profile, the compensation, e.g. matching the signals with the geometric data is performed.

To achieve the core part of RDLT, the referencing stage should be followed by the correcting process. The first step is to correct the data gained from Notch A as if the data on the time domain is obtained from an isotropic steel like a calibration block. Using the geometric data, Notch B is also corrected and the corresponding focal laws can be then created by cross correlation method which is described in detail in Chapter 5. The delay laws calculated are used to ensure that the beams emanating from each array element arrive at the time corrected aiming at a position where constructive interference occurs. For the validation purpose, the data accordingly corrected are parameterized in the phased array ultrasonic testing system. Then the improvements of SNR and locating flaws in original position with higher resolution are demonstrated.

4.3.3 Feasibility study by using austenitic calibration block (K1)

For the validation of the idea, the austenitic calibration block (K1) was employed. The sound velocity of the block is independent from the scanning points, e.g. identical velocity over every scanning point. Such a calibration block in ultrasonic testing is used for a "user calibration" of the equipment because ultrasonic equipment can be reconfigured for use in a large variety of applications. The "user calibration" includes the setting of the equipment, the configuration of the transducer and the testing setup for validation of the precision and accuracy. Hence, the K1 block is an ideal testing specimen to validate FMC-TFM combination technique since the measurement of the sound velocity by the technique is reliable. The block has the calibrated sound velocity 5,749 m/sec for the longitudinal and 3,135 m/sec for the shear wave. The geometric information is provided in Figure 4.7.

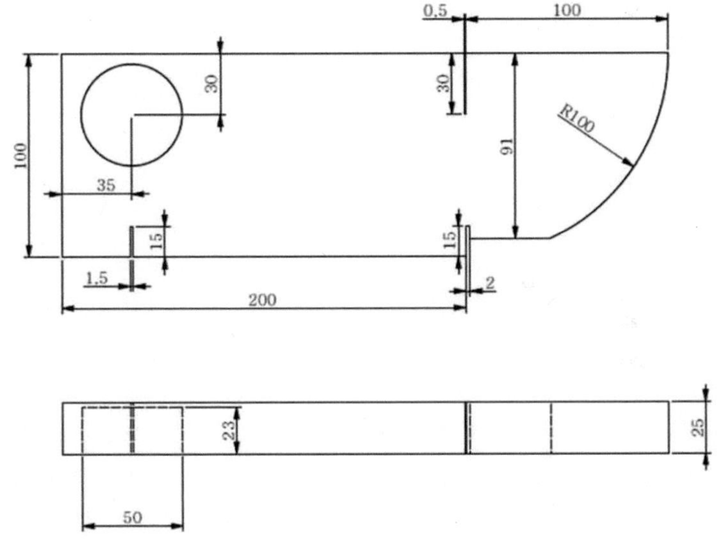

Figure 4.7: Geometry of austenitic calibration block K1

Figure 4.8(a) shows the measurement setup with an automatic scanning system. A linear phased array transducer with 16 elements, average center frequency 1.65 MHz and pitch size of 2.8mm was employed. The given pitch size is determined to improve signal-to-noise ratio (SNR). When all elements move in pitch size steps, e.g. the same position being transmitted as the position to be received, the SNR can be enhanced and the signal processing becomes simpler.

Figure 4.8(b) represents a schematic scanning mechanism by the advantage of the FMC-TFM combination technique focusing all the points in the ROI.

Once the total data set was obtained by FMC, two different post processing methods were then performed. First method is SAFT that is to measure the sound velocities from every scanning point (from P_1 to P_{46}). Therefore, the total number of velocity values gained is the same number as the scanning points.

Since there is a large reflector of 50mm in diameter just below the scanning range from P_1 to P_{15}, the signals acquired from this range were not chosen for the reason of precise sound velocity measurement. Hence, the signals reflected from the bottom P_{16} to P_{46} were selected. By using the elapsed time data fired from each element and reflected back from the bottom and the pre-known distance of the wave propagation at 31 numbers of scan points the sound velocity values were obtained and the whole data is provided in Appendix A.3. The graph representing the overall distribution of the velocity is shown in Figure 4.9. According to the measurements, the lowest velocity is 5,740.9m/sec, the highest one is 5,755.4 m/sec, the average velocity is 5,748.6m/sec and the standard deviation is 4.0m/sec. Taking the original velocity value of the calibration block, e.g. 5,749.0 m/sec into consideration, the FMC-TFM combination technique can provide a reliable sound velocity data. Hence, it is turned out that such an approach can be used as a tool for RDLT to determine the focal laws and to compensate the misplacement.

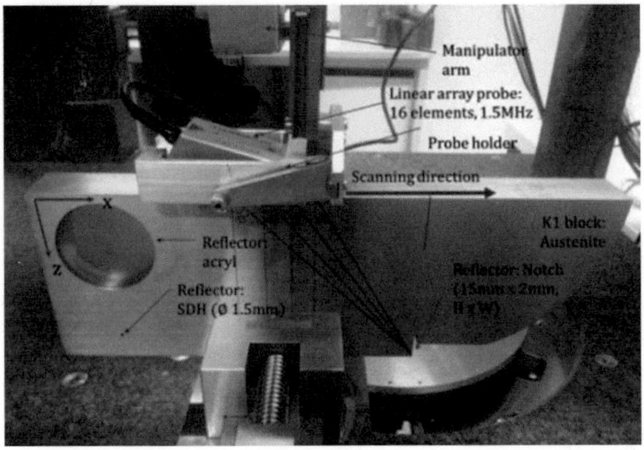

(a)

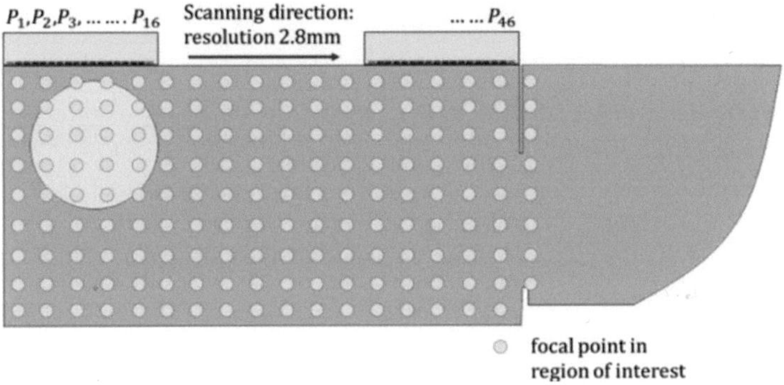

(b)

Figure 4.8: An image of the measurement setup for an austenitic calibration block (K1) by Referencing Delay Law Technique (a) and a schematic diagram of scanning with focal points in the whole region of interest (b)

The second post-processing method is TFM which reconstructs the image. As shown in Figure 4.3(a), the TFM is able to reconstruct desired points with complete data set gained from FMC. Figure 4.8(b) shows the focusing scheme for the calibration block aiming at in principle all the reflectors in the ROI including a notch tip in bottom.

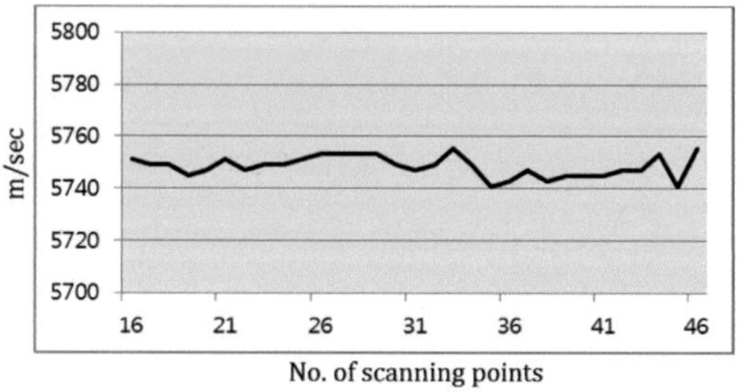

Figure 4.9: A graph of overall distribution of sound velocity gained from scanning point 16 to 46 by FMC-TFM combination technique

Figure 4.10 represents the results of the image reconstruction by TFM translucently covered on the photo copied image of the calibration block in 1 to 1 scale ratio. By comparing the both images with different values of SNR according to the equation below:

$$\mathrm{SNR} = 20\log_{10}\left(\frac{A_{flaw}}{A_{noise}}\right) \tag{4.8}$$

where A_{flaw} is the maximum amplitude of the flaw and the A_{noise} is the maximum amplitude of the noise signal.

Through this feasibility study, it was turned out that the proposed technique can provide the possibilities both of measuring the direct sound velocities at scan points which can be used to determine the focal laws and of improving SNR with the higher sensitivity for locating the notch-tip. Moreover, it was noticed that FMC-TFM can locate the minor signal of the side drilled hole in 1.5mm diameter which is challenging to detect. To achieve higher probability of detection (POD), in general, a discontinuity should be larger than one-half of the wavelength [72]. Since the half wave length of the transducer is about 1.92mm and the size of the hole is smaller, the results show a promising possibility to improve POD of minor size flaws in the austenitic welds.

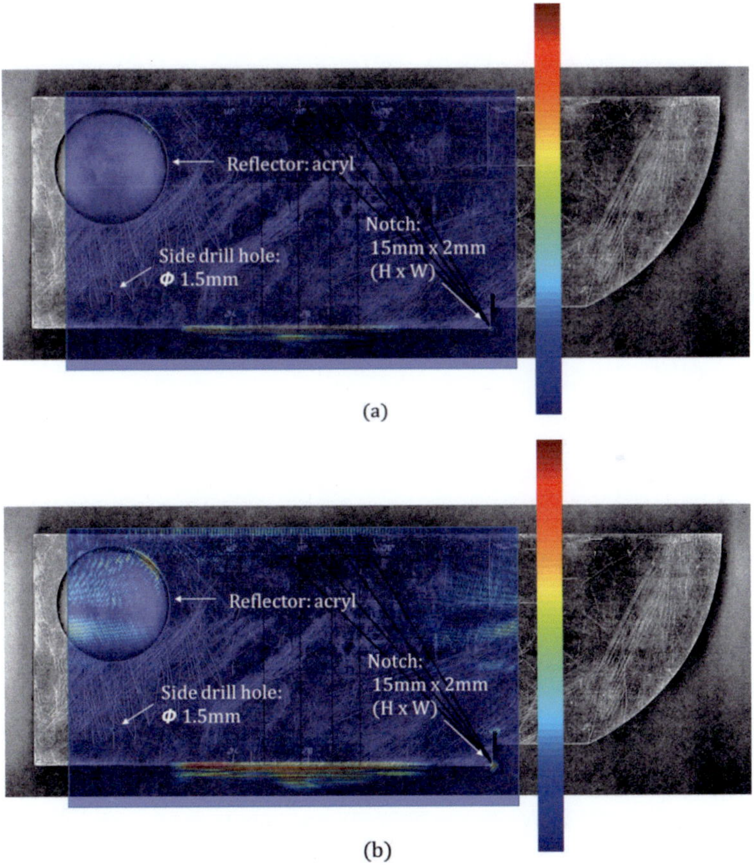

Figure 4.10: Post-processed images by Total Focusing Method on the austenitic calibration block: (a) signal-to-noise ratio of 3 to 1 (10dB) (b) signal-to-noise ratio of 9 to 1 (20dB)

Chapter 5.
Measurement by referencing delay law technique

5.1 Introduction

Once the specimen is mechanically prepared and is geometrically and crystallographically profiled according to Chapter 4, a relevant instrumentation should be followed. To achieve a precise and reliable testing, in particular electronic unit based measurement case, an optimized instrumentation should be selected. For an ultrasonic testing, the reliability of the testing results depends on the choice of transducer and equipment considering the characteristics of the test specimen. Hence, testing frequency, sensitivity, resolution, grating lobes, and SNR are the typical key factors in conventional ultrasonic testing to be considered. Based on such key factors, many transducer modeling theories, methods, and simulation programs have been introduced [73, 74, 75] and the studies for austenitic welds have been reported [76, 77, 78].

In this chapter, the RDLT was performed on the DMW specimen. Sound velocity measurements directly calculated were accomplished with the technique explained in Chapter 4 and they were considered as basic data. The data is then used for referencing the geometrical compensation and correcting the misplaced signals in sequence. The time delay values calculated by the RDLT were then utilized for validation by using phased array ultrasonic testing.

5.2 Preparation

5.2.1 Profile of ultrasound transducer

To obtain higher POD, the testing frequency must be well selected so that a discontinuity should be larger than one-half of the wavelength [72]. The typical rule of selecting a wave mode is stated that longitudinal wave becomes dominant due to the high attenuation and scattering associated with the use of shear waves especially on the austenitic welds [79].

The relationship between the ultrasonic wavelength and the grain size determines the scattering effects [80]. Scattering is critical, for instance, if the wavelength of the material has a similar dimension to the grain size. In case of the austenitic welds which have relatively big grain size, it is desirable to use a low frequency transducer to increase the wavelength. This, however, reduces the resolution.

At this experiment, a linear phased array transducer with 16 elements which has an average center frequency 1.65 MHz with a longitudinal wavelength of ~3.7mm was chosen and in the simulation study the transducer with the same specification was investigated in chapter 3.3.4. In order to identify the activated elements, scanning laser vibrometry (SLV) was performed on the transducer during polarization onto the elements.

The left photo in Figure 5.1 shows the bottom surface of the transducer and the image on the right represents the structure of the transducer during the optical activation. Since SLV requires a highly reflected surface condition of the sample, 3M™ Scotchlite™ High Gain Reflective Sheeting 7610 was sealed on the surface of the PZT of the transducer. For the reason of protecting 1st and 16th element, two elements are designed at each outside and they are functionally inactivated. Each element was set to be fired in the different time interval and it was revealed in odd- and even number of the elements. Table 6.1 represents the key parameters for the phased array testing.

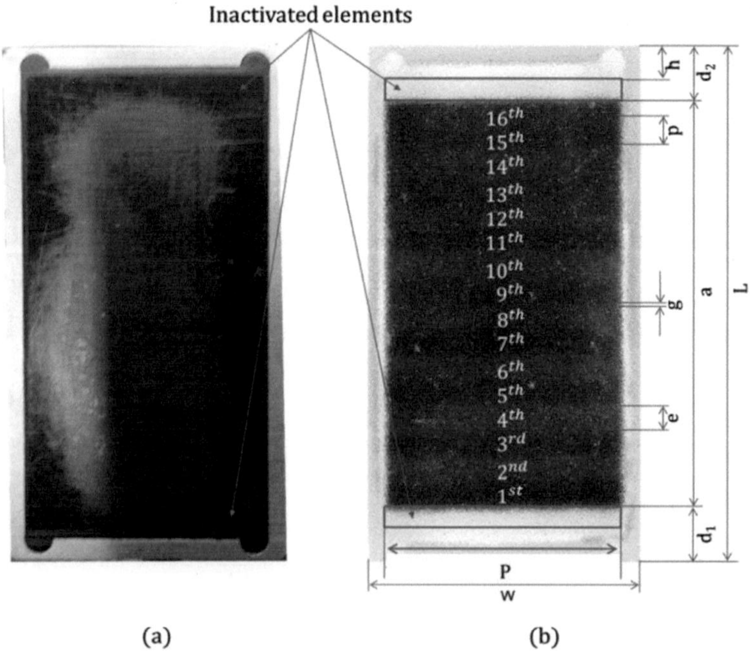

Figure 5.1: Identification of a linear phased array transducer using scanning laser vibrometry (SLV) (a) bottom surface of real transducer (b) image cut during polarization measured by SLV

[unit: mm]

Nomenclature	L	a	W	P	h
Parameter	length	active aperture	width	passive aperture	housing depth
Dimension	58.00	44.80	31.80	26.00	3.80
Nomenclature	d_1	d_2	P	g	e
Parameter	distance housing to 1st element	distance housing to 16th element	pitch	gap between two adjacent elements	element width
Dimension	6.60	6.60	2.80	0.08	2.72

Table 6.1: Values of general dimension and key parameters of linear phased array transducer used in experiment

5.2.2 Introduction to PCUS Pro® array and requirements for instrumentation

Both FMC-TFM and FMC-SAFT combination technique have a key role on this dissertation. To apply the technique, a phased array ultrasonic testing system should be optimized for processing fast data transfer from the array controller unit to the storage unit. In order to test the heavy-wall DMW specimen, in addition to the selection of low frequency transducer, the pulser-receiver must have the bandwidth optimized to use the low frequency transducers and the resolution of the time delay should be 1 ns between the elements being fired. It is typically necessary to use filters so as to process the signals actively in the receiving part. Therefore, it was determined to use the PCUS Pro® phased array ultrasonic testing equipment as shown in Figure 5.2 [81].

Figure 5.2: PCUS Pro® Array electronics (Fraunhofer IZFP-D)

5.2.3 Measurement setup

An automated scanning ultrasonic testing system with a highly stable working desk was used. Due to the complicated shape of the specimen, an angle vise was used for positioning the specimen horizontally flat. The geometry of the specimen can be specified as an input file that can be read by the PCUS software before starting the measurement. Figure 5.3 shows the automatic measurement setup.

The scanning step is 2.80mm corresponding to the size of the pitch and the total number of scanning steps is 50 for covering the whole area of the weld. For the coupling between the transducer and the specimen, ULTRAGEL II ® (Sonotech, IL, USA) was used.

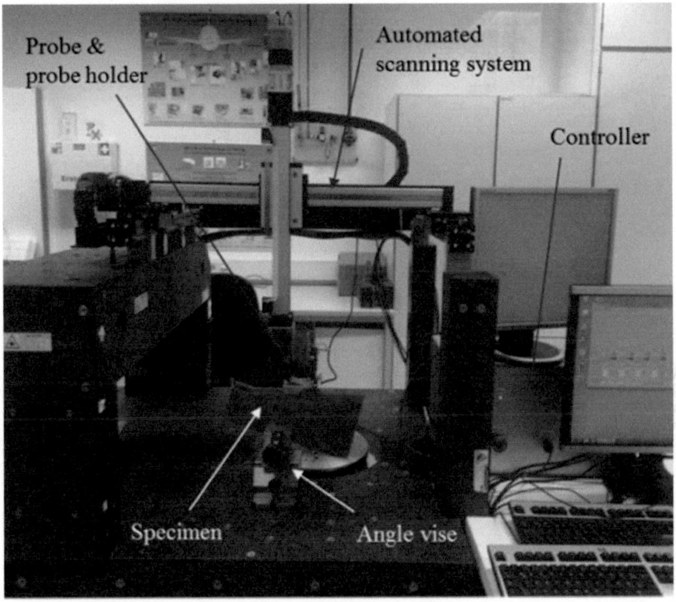

Figure 5.3: An image of measurement setup at the reference block for Referencing Delay Law Technique (Fraunhofer IZFP-D)

5.3 Data acquisition

5.3.1 Sectorial scan mode

In the S-scan measurements, the sound velocity was set to 5,718 m/sec. For the configuration of the pulser, the pulse length is 225ns and the sender voltage was set to 80volts. The sampling rate was 80MHz for recording. The sweep angle range was 15 to 31 degree for Notch A and 0 to 25 degree for Notch B as shown in Figure 5.4. The both sweep angles have the resolution of 0.1 degree to acquire a higher lateral resolution. The scanning resolution was set to 2.8mm corresponding to the pitch size of the transducer and the scanning scope is from P_1 to P_{50} for covering whole ROI.

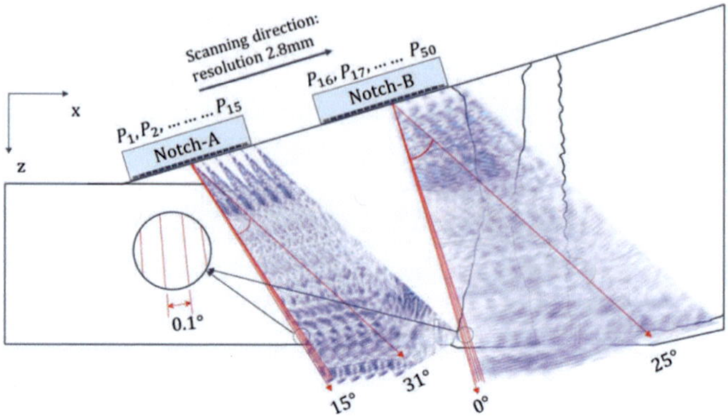

Figure 5.4: An image of a schematic diagram of sectorial scan

5.3.2 Full matrix capture

In the FMC-TFM measurements, pulser configuration, sender volts and sampling rate were set in the same condition as the sectorial scan. The total number of steps of the scan is 50 with the scanning resolution of the pitch size, of which the condition is the same as the S-scan measurements.

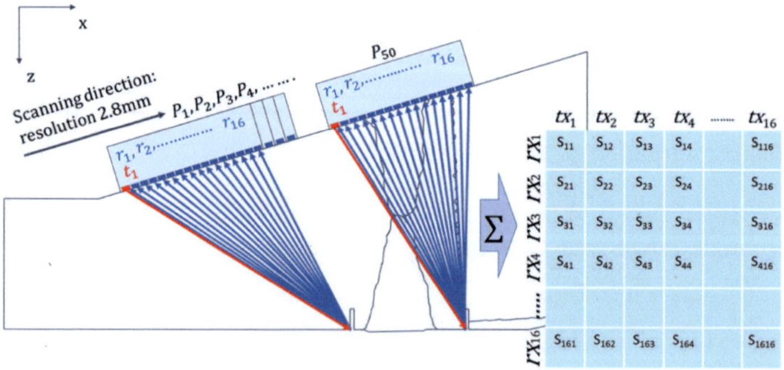

Figure 5.5: An image of a schematic diagram of scanning with focal points in the whole region of interest

Figure 5.5 represents the schematic diagram of the scanning by FMC-TFM focusing points at the whole region of interest. The number of sampling points is 5,600. By multiplying the number of element of transmitters and receivers, the scanning points and the sampling points, the total number of the RF data acquired is 71,680,000. Such the number of data is rather than sufficient to utilize for the RDLT

5.4 Data analysis and identification of degradation by S-scan mode

As a coordinate of the signal acquired from the PCUS Pro®, both the beam propagation distance from the activating element to the sound reflector and the angle received from the sound reflector within the steered angle are displayed on an S-scan image and recorded in the software of the PCUS Pro®.

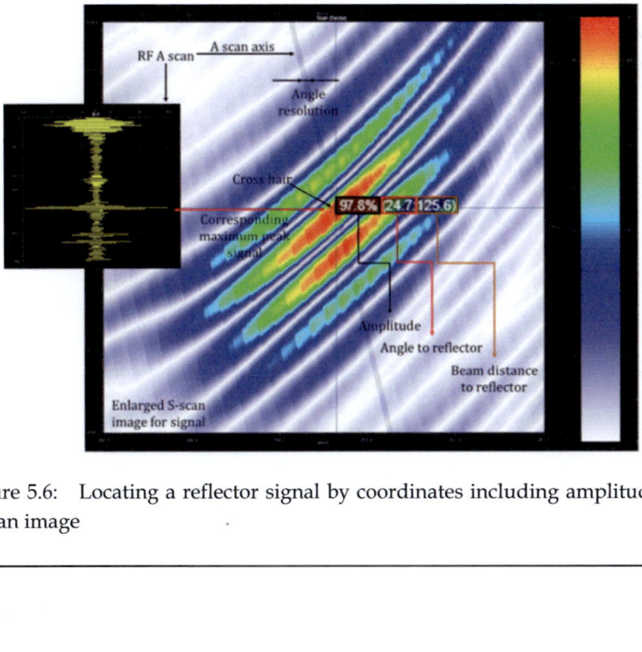

Figure 5.6: Locating a reflector signal by coordinates including amplitude in S-scan image

As shown in the Figure 5.6, in order to find a maximum signal amplitude in the enlarged S-scan image, an axis of beam representing a defined resolution angle which corresponds to an A-scan can be chosen. When a cross hair fits with the maximum peak, the amplitude in percent, beam propagation distance to the flaw, d_F, and steered angle to the flaw, θ_F, are indicated which are used for locating a flaw precisely. Based on the acquired coordinates from the PcusPro, the data can be transformed into (x, z) coordinates of the geometrical structure as shown in Figure 5.7 by the simple equation below:

$$x = d_F \cos\theta \qquad z = d_F \sin\theta \tag{5.1}$$

where d_F is the wave propagation distance to the focal point

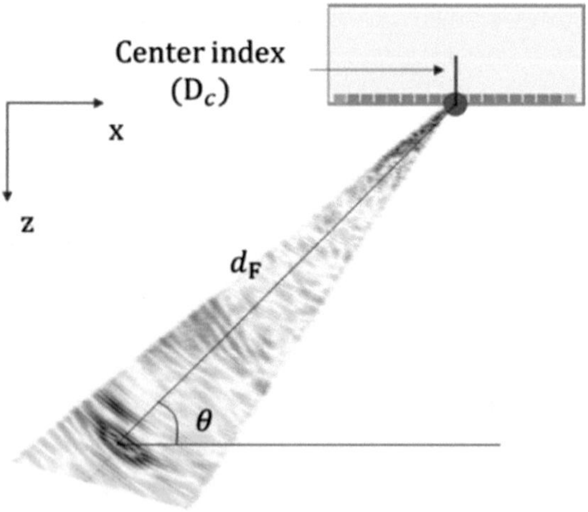

Figure 5.7: Schematic diagram of transforming coordinates by S-scan into (x, z) coordinates

Among the scan position from P_1 to P_{50}, the maximum amplitude signals with the highest sensitivity and resolution to the Notch A and Notch B are found at P_8 and P_{30}, respectively. The magnified images represent the signals corresponding to the signals detected. All the S-scan images which are covered on the photo of the reference block in this Chapter are fitted with the actual size photo in 1 to 1 scale.

Figure 5.8 shows S-scan images measured at scan position 8 aiming at Notch A. The signals highlighted in a square line in the S-scan images are identified as shown in Figure 5.6. The wave propagation distance to the root-edge of Notch A and the angle coordinate 0 degree angle to the root-edge are then compared to the geometric data. In the case of the root-edge signal in Notch A, the signal was found behind Notch A in 2.5mm distance which is mainly assumed that the sound velocity inserted into the phased array would be different from the velocity of the media where the wave travels.

The signals gained at Notch B are significantly misplaced not only in horizontal but in vertical direction which would represent strong anisotropy in the weld. Apart from the artificial flaws, a signal reflected inside the weld which is suspected as an inclusion or weld defect is found in the middle area of the weld as shown in Figure 5.9. This suspected signal has not been successfully investigated yet. The signal was not detected by the immersion ultrasonic testing in B- and C-scan images as shown in Figure 3.22. It was assumed that the size of the sound reflector is too small to be found by the ultrasonic testing transducer. Taking one half the wave length of the transducer in the immersion testing, e.g. around 1.4mm, into consideration, the suspected reflector would be less than half of the wave length [72]. Beside of the immersion testing, Digital X-ray CT was performed to identify it but due to the strong absorption of the weld, the CT failed. Such a small size inside reflector in heavy wall austenitic weld can be identified by a gamma-ray CT. Since this kind of small inner sound reflector is often found in the weld and is not classified as an essential defect which must be detected according to related standard, i.e. ASME code, the signal was just monitored before and after the image processing by RDLT. The results after the processing were described in Chapter 5.6.

The longitudinal waves are mode converted at an interface. However, in this dissertation only the longitudinal wave mode was considered. In case of S-scan image in this study, the mode converted and refracted shear and longitudinal waves are found but the effects are not considerable. In terms of mode conversion, FMC-TFM combination technique has the great advantage of avoiding such a complicated signal analysis.

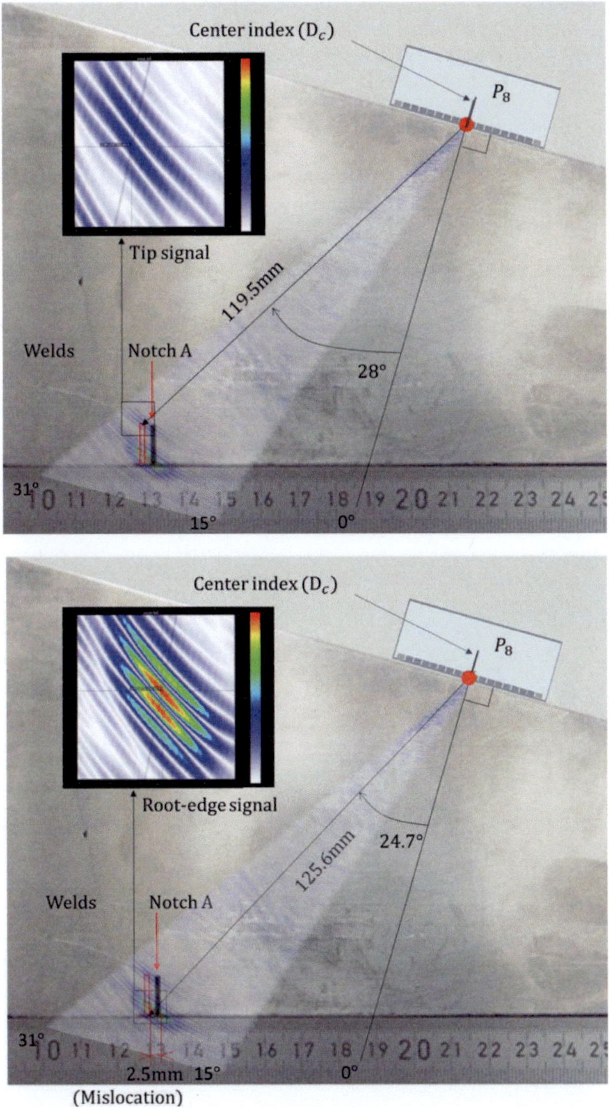

Figure 5.8: S-scan images scanned across Notch A representing misplacement before compensation of time delay laws, focusing at notch-tip (top) and root-edge of notch (bottom)

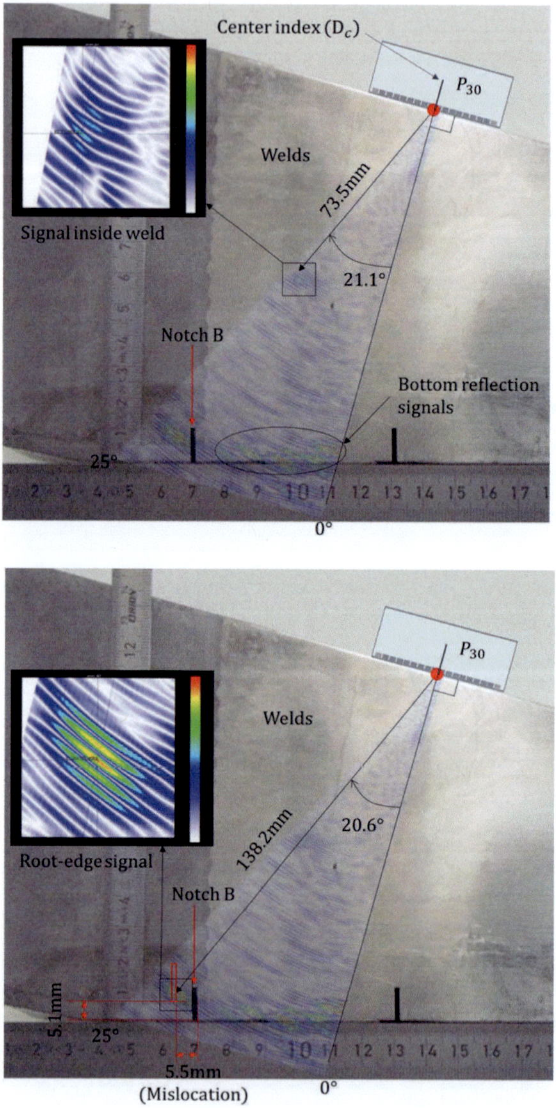

Figure 5.9: S-scan images scanned across Notch B representing misplacement before compensation of time delay laws, signal inside weld unfocused (top) and focusing at root-edge of notch (bottom)

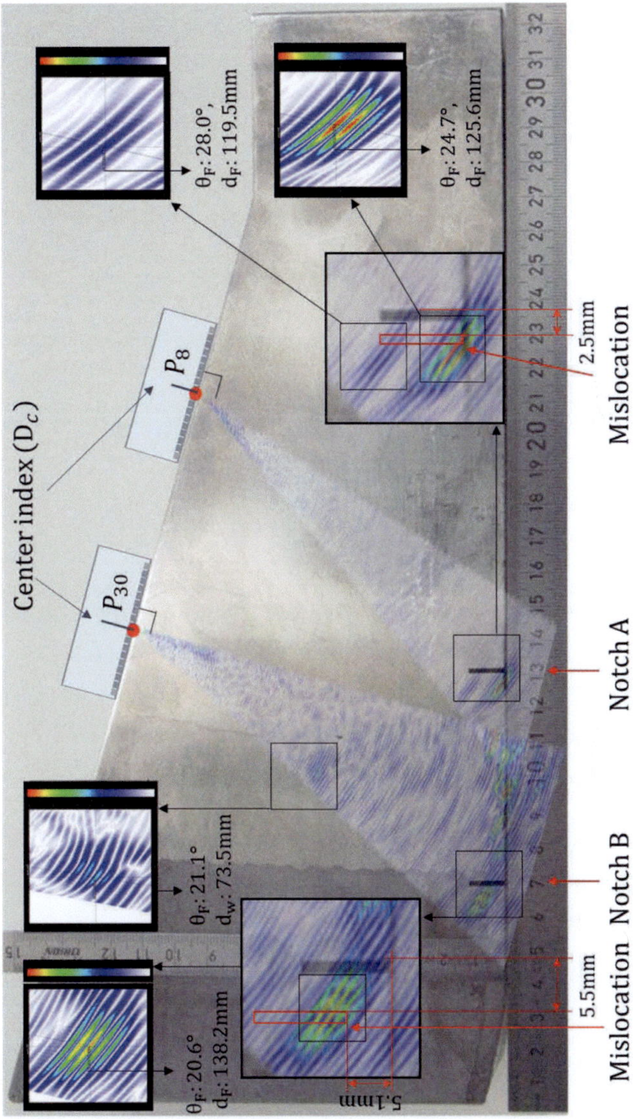

Figure 5.10: Sectorial scan images scanned across Notch A and B before compensation of time delay laws

5.5 Referencing and correcting process

5.5.1 Sound velocity measurements by FMC-SAFT combination technique

In order to process RDLT, it is a key approach that the transversely isotropic region, in terms of the elastic properties in different grain direction, should be corrected. For this, it is important to know the sound velocity in horizontal and in vertical direction of the wave path. By using FMC-SAFT the sound velocity values can be directly determined.

5.5.1.1 Sound velocity measurements for 0 degree angle in Z-axis (c_Z)

As described for the direct velocity measurement using FMC-SAFT combination technique in Chapter 3, the values of sound velocity in 0 degree in Z-axis were acquired based on the basic data set gained from scanning point P_1 to P_{35}. Due to the weld area where a homogeneous sound velocity measure can hardly be achieved, the data set obtained from P_{36} to P_{50} was not considered. Figure 5.11 shows the schematic diagram of the scanning method.

Since every scanning step is 2.8mm, i.e. pitch size , the distance in 0 degree angle in Z-axis at each step d_x is increased by $2.8 \times \sin 16$ as shown in Figure 5.12. After scanning is performed from the scanning point P_1 to P_{35}, the signals are processed by SAFT and the time arrival data gained from each scanning point is used for sound velocity measurements.

In addition to the time arrival data, utilizing the pre-known data of the depth distance of the first scan point d_{x0}, the velocity measured at P_1 can be obtained. Since the distance of movement according to the pitch size and the inclined angle are also available, using the Pythagoras' theorem, i.e. $2.8 \times \sin 16$

the depth distance of the rest scan points are determined and used for calculating the sound velocity in Z-axis. Hence, the sound velocity at certain point can be determined. Finally, after obtaining all the arrival time values from P_1 to P_{35}, linear fit was performed as shown in Figure 5.13.

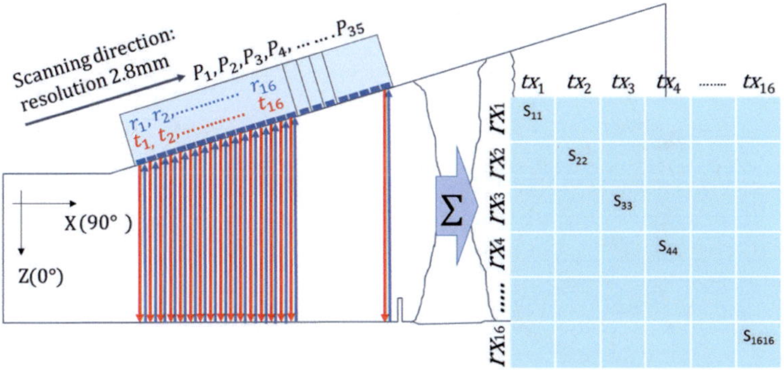

Figure 5.11: An image of a schematic diagram of scanning for measurements of direct sound velocity in 0 degree using FMC-SAFT combination technique

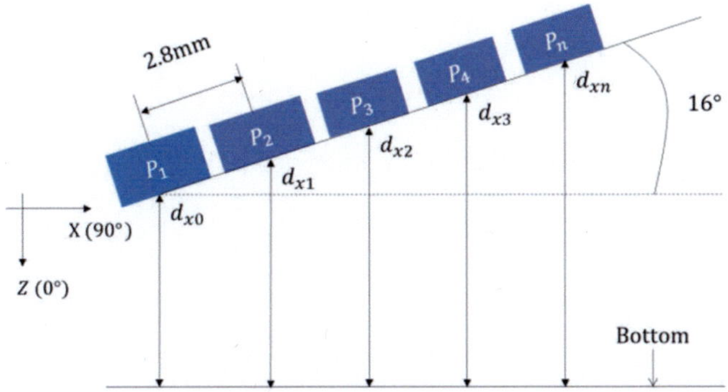

Figure 5.12: A schematic diagram of determination of depth distance in 0 degree angle in Z-axis at each scanning point

The graph related to the overall distribution of the velocity from each scanning point is shown in Figure 5.14. According to the measurements, the lowest velocity is 5,606m/sec, the highest one is 5,618.1m/sec, the average velocity is 5,612.6m/sec and the standard deviation is 3.4m/sec. The 35 values of time, depth and sound velocity are provided in Appendix A.4.

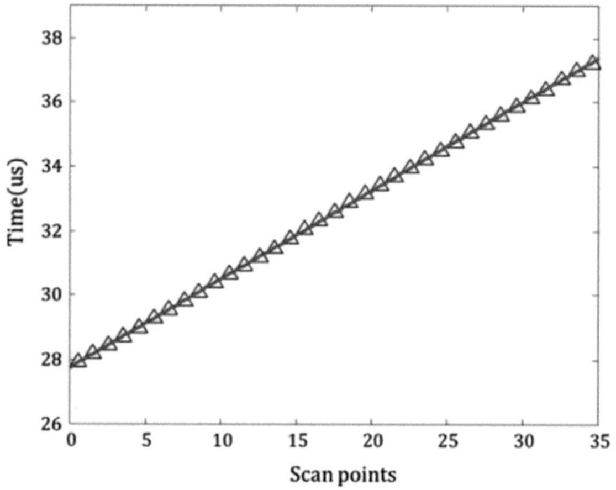

Figure 5.13: A graph of time distribution for scan points by linear fit

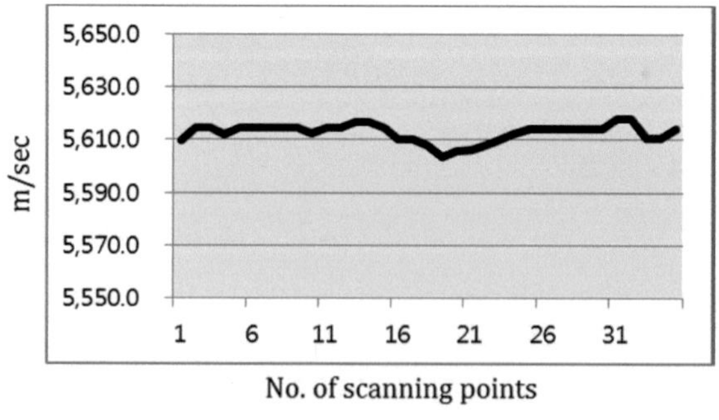

Figure 5.14: A graph of overall distribution of sound velocity gained from scanning point 1 to 35 in 0 degree angle (Z-axis)

5.5.1.2 Sound velocity measurements for 90 degree angle in X-axis (c_X) and compensation of error in transversely isotropic region

Now the sound velocity data in Z-axis is reliable. To obtain the sound velocity value in X-axis by RDLT, a geometric based calculation using the signal data processed by SAFT was performed. At the velocity measurement of Z-axis, it was intended to obtain a height of Notch A with maximum amplitude. It was assumed that the wave path direction is perpendicular to the bottom line as shown in Figure 5.15.

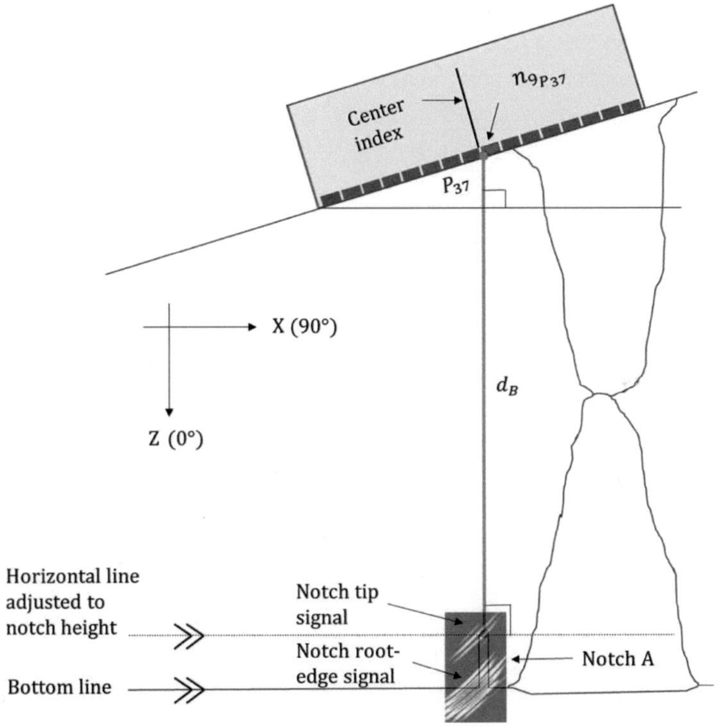

Figure 5.15: Schematic diagram of gaining maximum amplitude signal reflected from Notch A, $n_{9_{P_{37}}}$ - the 9th element at scan position 37, 9th element in red – activation in transmitting and receiving

Under this scheme, the transducer was placed at the scan position 37 (P_{37}), by using the center index of the linear array transducer where only the 9th element was activated for signal transmitting and receiving. By analyzing the signals processed by SAFT, the height of the notch was precisely identified and the horizontal line indicating the height of the notch was used for adjusting the beam paths gained from other scan positions. The image of notch tip and notch root-edge signals in Figure 5.15 helps to understand the method. The image was partially cutout from the reconstructed image of notch A by FMC-TFM.

Figure 5.16 illustrates the schematic principle how to determine the sound velocity in X-axis. In the figure the scan position moved onto P_{20} and due to the geometric dimension the distance of beam path became longer than the one at P_{37}. Based on the scanning path from P_{37} to P_{20}, the moving distance of the transducer in horizontal line is defined as X_I. d_I lined in red is the distance of the beam path to be gained as an ideal measurement, e.g. in case of the sound velocity is fitted with the beam path. The beam path in blue marked as d_R is the reconstructed image providing the image of the notch tip and root-edge signals. It was assumed that when the wave (d_B) passes into the specimen in Z-axis at P_{37}, there would be no component of the sound velocity in X-direction. In the same way, the beam (d_R) generated at P_{20} would arrive at the reflector with the sound velocity in Z-axis, e.g. c_Z. Then, the time of flight of d_I is the same as d_R of which each beam arrives at the same height of the notch, but only the sound velocity of d_I, e.g. c_I, and d_R, e.g. c_R, is different as shown in Figure 5.17.

In order to obtain the value of c_X, it was assumed that when a reconstruction is processed by the value of c_Z, the signals of Notch A should appear in front of its real position in space as shown in Figure 5.16. As a result, the difference in distance between d_I and d_R, e.g. ΔX, becomes the function for calculation of the sound velocity in X-axis.

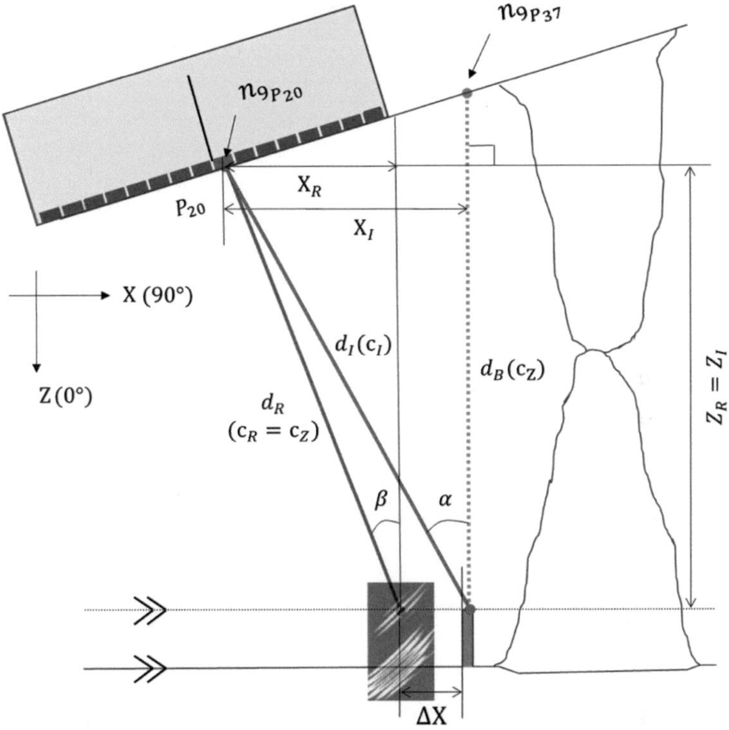

Figure 5.16: Schematic diagram of determining ΔX to calculate sound velocity in X-axis using signals reflected from Notch A at scan position 20. d_I – ideal beam path, d_R -beam path reconstructed by the velocity in Z-axis.

According to the scheme in Figure 5.16, d_I must travel into the media with the sound velocity c_I which is defined as following:

$$\vec{c}_I = c_X \cdot \sin\alpha \cdot \vec{u}_X + c_Z \cdot \cos\alpha \cdot \vec{u}_Z$$

$$c_I = |\vec{c}_I| = \sqrt{c_X{}^2 \cdot \sin^2\alpha + c_Z{}^2 \cdot \cos^2\alpha} \tag{5.2}$$

For the image reconstruction, the same processing to determine the sound velocity of c_R as an ideal beam condition (c_I) is performed. Under the condition of $c_X = c_Z$, the sound velocity is also assumed as below:

$$c_R = |\vec{c}_R| = c_Z$$

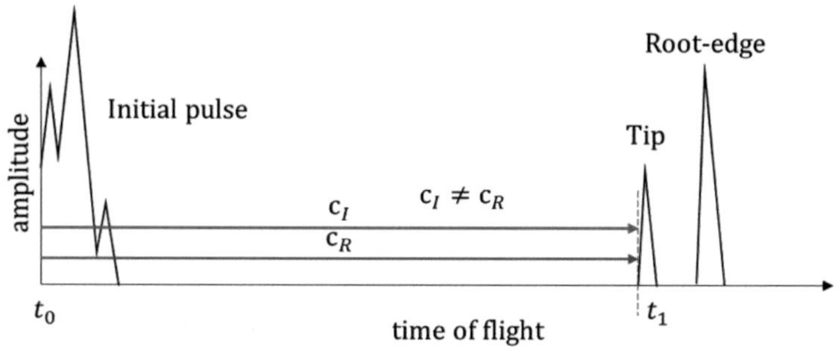

Figure 5.17: Schematic diagram of relationship between sound velocity of ideal beam (c_I) and reconstructed signal (c_R) of Notch A based on time of flight

Then, by using the physical data as shown in Figure 5.17, the difference in distance in X-axis between the wave path at P_{37} and P_{20}, e.g. X_I is defined as follows:

$$X_I = X_R + \Delta X$$

Each of X_I and X_R is then determined as follows:

$$X_I = c_X \cdot \sin\alpha \cdot t_1$$

$$X_R = c_Z \cdot \sin\beta \cdot t_1$$

Hence, the ΔX is finally defined as following:

$$\Delta X = X_I - X_R = (c_X - c_Z) \cdot \frac{X_R}{c_Z} \tag{5.3}$$

Since we know the coordinates of Notch A from the geometric data, by using such a data ΔX is determined and hence the sound velocity of c_X at a certain scan position is calculated by the following equation:

$$c_X = c_Z \frac{X_I}{X_I - \Delta X} \tag{5.4}$$

When the beam path for gaining ΔX is generated near the scan position, e.g. P_{37}, it would be relatively hard to determine the ΔX because the difference is too slight to distinguish. Due to the physical nature with the beam path distance, e.g. the longer the X_R value, the bigger the ΔX as defined in Eq. 5.3. Hence, the calculation for c_X would be reliable when the measurements are performed at the scan position from P_{16} to P_{28} representing 13 individual data.

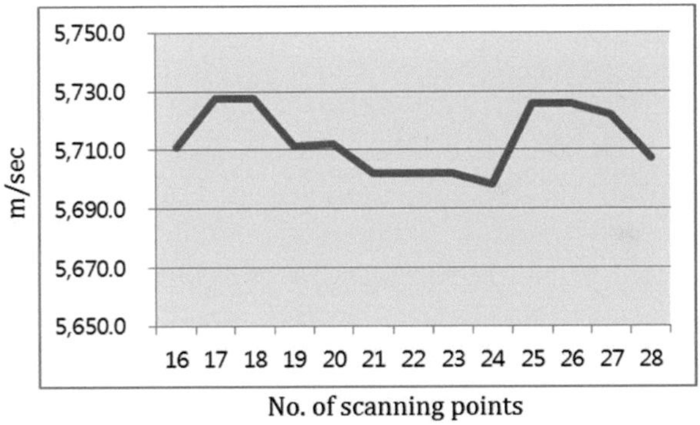

Figure 5.18: A graph of overall distribution of sound velocity gained from scanning point 16 to 28 in 90 degree angle (X-axis)

Through the calculation gained from each scan position from P_{16} to P_{28}, the individual sound velocity was determined and Figure 5.18 shows the distribution of the values. The standard deviation is 11.1 m/sec and the average velocity is 5,713.4m/sec. Every velocity value is listed in Appendix A.5.

5.5.1.3 TFM processing for adjusting Z-axis based on Notch A

When the sound velocity values of the material in X- and Z-axis are determined, correcting the sound velocities in X- and Z-axis in the transversely isotropic region becomes the first step. For the image reconstruction by TFM, the average sound velocity of the transversely isotropic region in X-axis was used as input. Then the images were adjusted to the Z-axis. As a result, the signals of Notch A are located as shown in Figure 5.19. A reference line indicating the center of Notch A is used for fitting the image for further processing although the root and tip signals of Notch B still remain out of position.

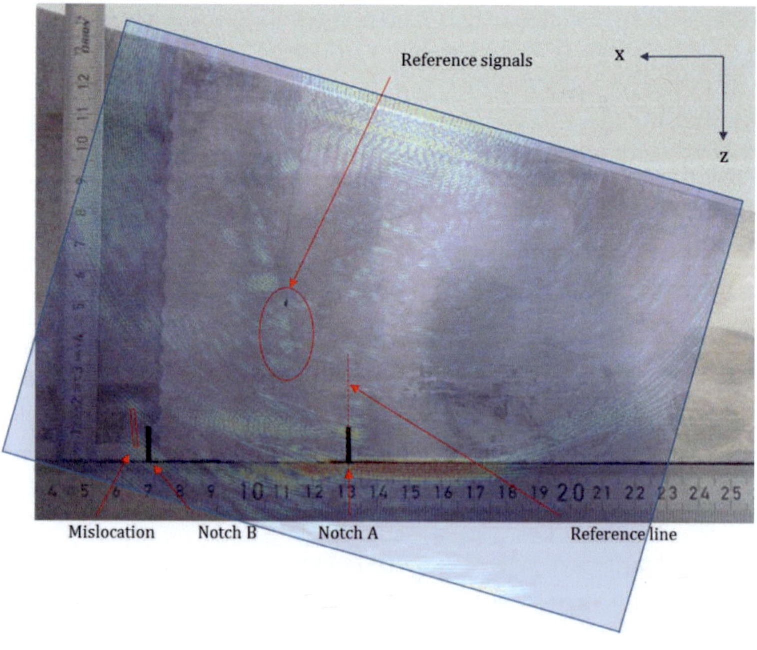

Figure 5.19: Image gained by total focusing method for Notch A and B

In addition, the scattered signals reflected strongly at the boundary of the weld in the middle lower region can be also used for adjusting the TFM image. It was noticed that the weld boundary in the upper area is not optimized for receiving relatively weak reflection signals like scattered signals. Mainly due to the strong absorption in the austenitic weld and the buttering area, scattered signals are hardly distinguished for reference purpose. Figure 5.20 shows the binary image of the scattered signals in black and white. In order to see clearly the boundary, the boundary of the weld was re-traced in black.

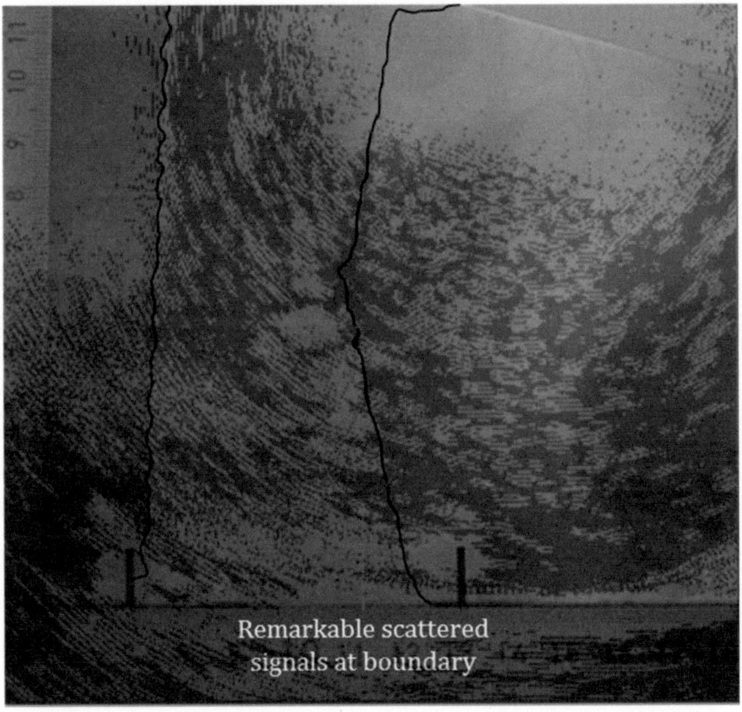

Figure 5.20: Binary image gained by total focusing method for validation of scattered signals mapped on metallography

5.5.1.4 TFM processing for adjusting Notch A in X-axis

Now the Z-axis in the transversely isotropic region has been fitted but the X-axis has not accomplished yet. Since the sound velocity was set to the value of c_X at the TFM processing for adjusting in Z-axis, the wave propagating in Z-axis arrives at the sound reflectors as much as the sound velocities in X- and Z-axis are different. As a result, Figure 5.21 shows the ultrasound echoes propagated longer than the real height of Notch A. For locating the signals gained by TFM onto the 1 to 1 scale image photo of the specimen, the image of the signals with maximum amplitude according to the spectral scale in right side were placed. Then the scale in the TFM image in left side and the scale (steel scale) on the specimen is used to place the image in correct position.

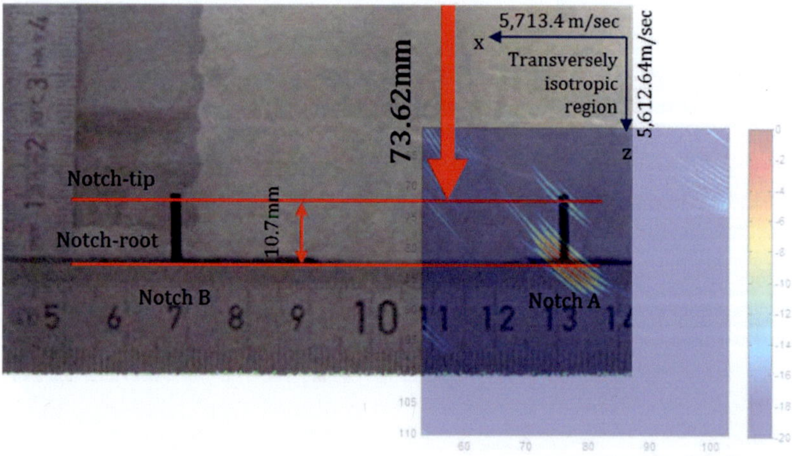

Figure 5.21: Image gained by TFM for Notch A representing missizing of the notch height before the compensation of the base material where the sound waves with the equal sound velocity of 5,713.4m/sec in x- and z-direction pass through the transversely isotropic region

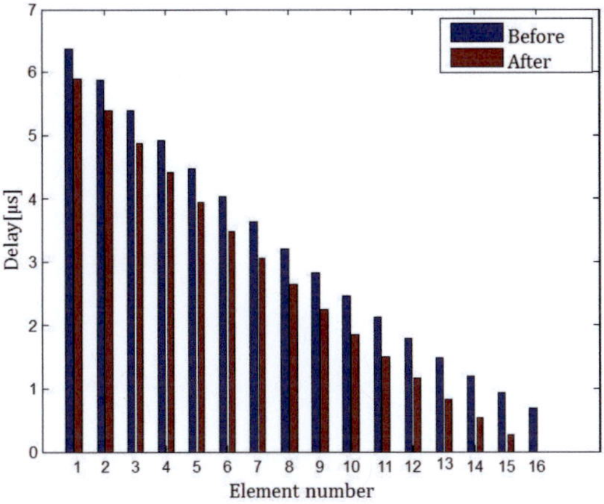

Element No.	Before correction [μs]	After correction [μs]
1	6.3750	5.9000
2	5.8750	5.3875
3	5.4000	4.8875
4	4.9250	4.4125
5	4.4750	3.9500
6	4.0375	3.5000
7	3.6250	3.0625
8	3.2125	2.6500
9	2.8375	2.2500
10	2.4750	1.8625
11	2.1250	1.5125
12	1.8000	1.1625
13	1.4875	0.8375
14	1.2000	0.5375
15	0.9250	0.2625
16	0.6875	0

Figure 5.22: Focal law values of each element at receiving for geometrical compensation of Notch A by the compensation of the base material where the sound waves pass through the transversely isotropic region

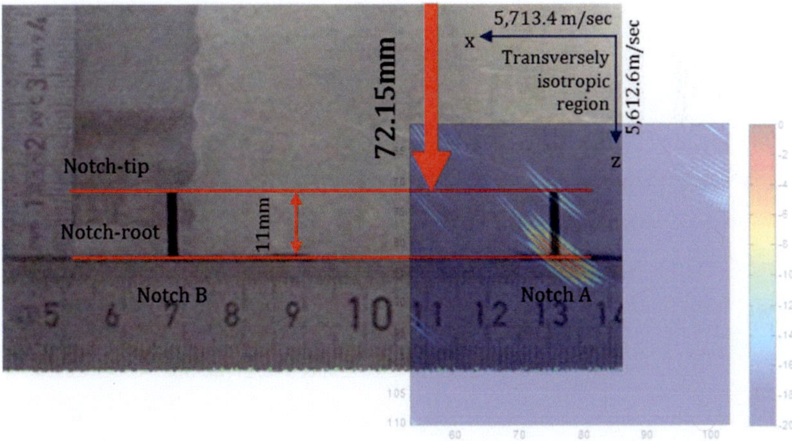

Figure 5.23: Image gained by TFM for Notch A after the compensation of the base material where the sound waves with 5,713.4m/sec in x-direction and with 5,612.6m/sec in z-direction pass through the transversely isotropic region

In order to fit the height of the notch, the time delay value of each element was calculated by using the sound velocity obtained in Chapter 5.5.1.2 and is shown in Figure 5.22. Now the time delay values in the transversely isotropic region are determined and then applied to compensate the error of the height of Notch A. Figure 5.23 represents the corresponding results of TFM gained at Notch A. The notch tip and the root-edge signals are well located in both horizontal and vertical direction.

5.5.1.5 Geometric based computation and compensation of error in austenitic weld region

Now the error in the transversely isotropic region is compensated but the austenitic weld region remains to be processed. When the sound velocity is simply different in the material, the velocity values can be obtained by the method applied for the transversely isotropic region. However, in the case of the austenitic weld, the wave skew should be considered. Taking the X-axis for the first step into consideration, Figure 5.24 shows that the height of Notch B is 0.8mm higher than the original dimension which is the opposite case against Notch A. Unlike the signals obtained from the transversely isotropic media, the signals gained through the anisotropic and heterogeneous media contain bottom-wall reflected signals due to the beam skew. In addition, at the image reconstruction it was interesting to notice that all the elements form the wave signal in a circular arc shape and it generates the noise like signals as shown in Figure 5.24. These complicated signals would hinder the evaluation and cause misinterpretation.

To determine the time delay between two different signals, the cross correlation method is used. In the method the time shift of the echoes is estimated directly from the RF-signal.

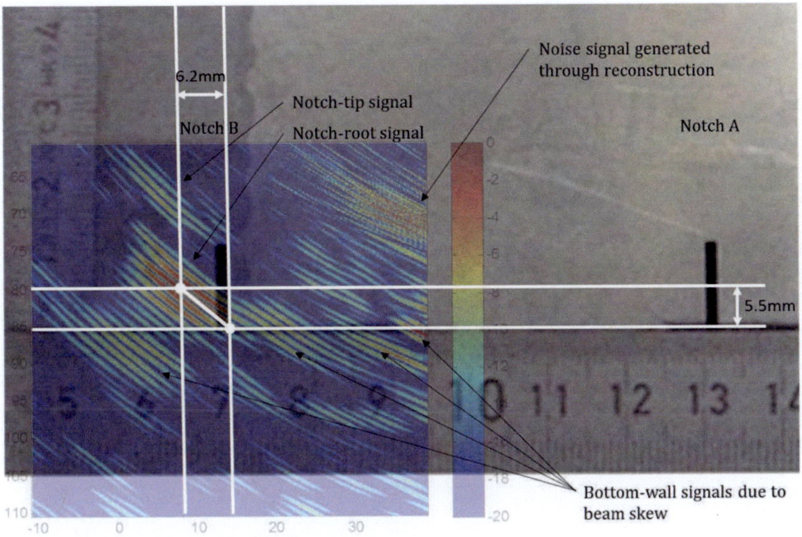

Figure 5.24: Image gained by TFM for Notch B representing missizing of the notch height before the compensation of the base material where the sound waves pass through the weld region by the geometrical computation

The time delay is determined by searching for the maximum correlation coefficient between the successive received echoes. The correlation function which is normalized by dividing the number of correlations T is defined as follows:

$$C(x, \delta t) := \frac{1}{T}\int_0^T u(x_{ref}, t)\tilde{u}(x, t - \delta t)\mathrm{d}t \tag{5.5}$$

where u is the received signals from the reference block, ũ is the ideal signal based on the geometric computation of the reference block, t is the elapsed time after wave reaches the target notch.

Then the correlation function has maximum magnitude which indicates the point in time domain and the time delay is defined as following:

$$T_{delay} = \max_t C(x, \delta t) \tag{5.6}$$

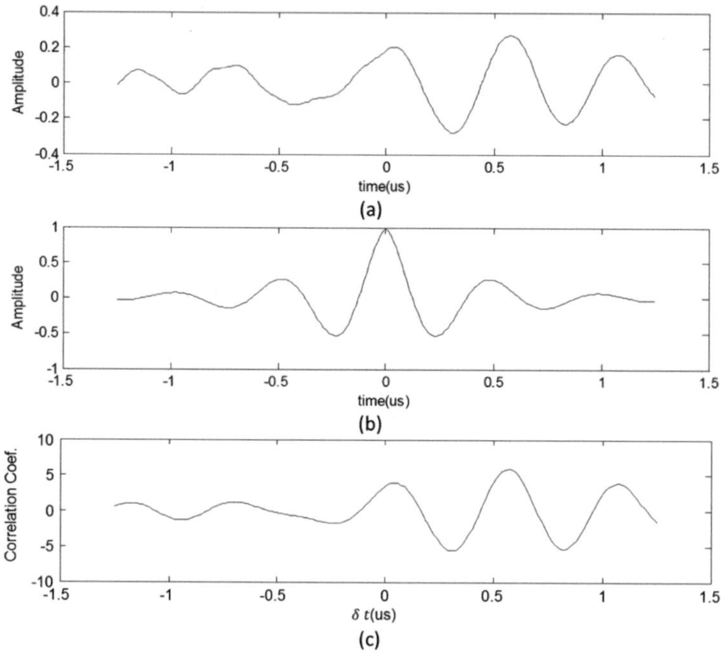

Figure 5.25: Illustration of an example of cross correlation between two different signals measured at 14th element at 38th scanning point , (a) signals obtained from reference block, (b) ideal signals based on geometry and (c) correlation coefficient with maximum time delay

Figure 5.25 illustrates three different signals as an example which is obtained at 14th element at P_{38} between the signals before and after geometry based compensation by corrected delay laws for Notch B. Figure 5.25 (a) shows the signals measured from the reference block, Figure 5.25 (b) represents the geometry based ideal signals where the peak echo is shown in zero at the time line and Figure 5.25 (c) is the correlation coefficient indicating the maximum time delay of 0.5875μs.

By using the pre-known dimension of Notch B, i.e. the height of the notch, the time delay values by using cross correlation method are calculated and the delay values for correcting the error of the height are shown in Figure 5.27 with green bars. Although the error of the height is compensated, the reconstruction result still shows 4.4mm misplacement from the original location in X-axis as shown in Figure 5.26.

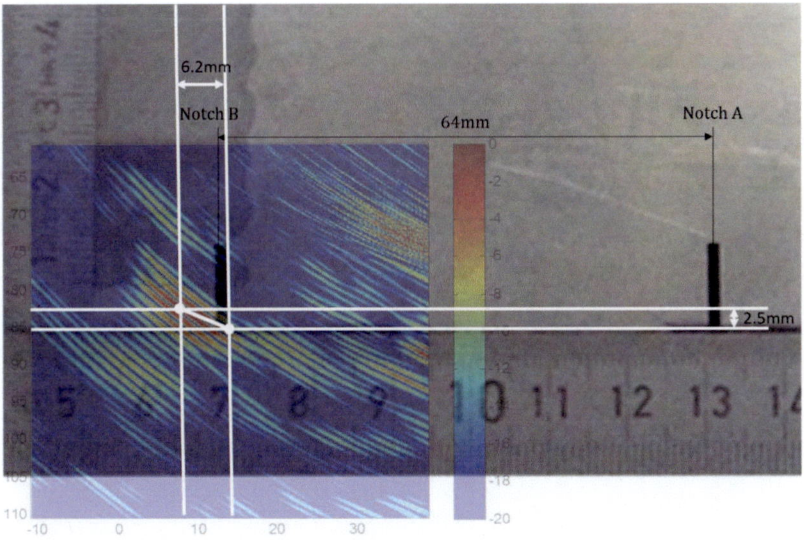

Figure 5.26: Image gained by TFM for Notch B representing misplacement of the notch in X-axis after the compensation of the base material where the sound waves pass through the weld region by the geometrical computation

This is mainly caused by the wave skew in the weld. Thanks to the geometric information, location of Notch B signal can be determined by using the distance between the center of Notch A and the center of Notch B in horizontal axis, i.e. 64mm. The time delay values can be determined by using cross correlation method. Hence, the delay laws can be determined by referring both the geometric location of Notch B and the location of the signal acquired by the cross correlation method. The time delay values acquired by this method for correcting the misplacement are provided in Figure 5.27 in red bar. By this RDLT, the signals of Notch B including the notch tip are more precisely located as shown in Figure 5.28.

Such a delay law gained from RDLT is used for correcting misplacement of signal and this can be also validated by phased array ultrasonic testing. That is the objective of this study to realize the quantification of the defects with higher POD in challenging materials, for example austenitic welds.

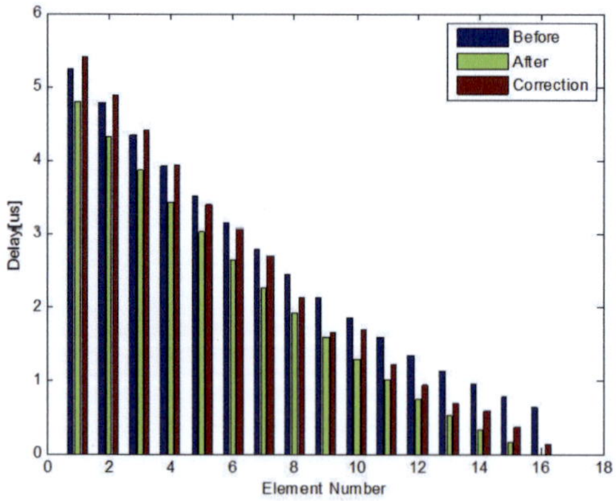

Element No.	Before correction [μs]	After correction [μs]	Delay correction [μs]
1	5.2375	4.8000	5.4125
2	4.7750	4.3250	4.8875
3	4.3375	3.8750	4.4125
4	3.9250	3.4375	3.9375
5	3.5250	3.0250	3.4000
6	3.1500	2.6375	3.0625
7	2.7875	2.2625	2.6875
8	2.4500	1.9125	2.1375
9	2.1375	1.5875	1.6500
10	1.8500	1.2875	1.6875
11	1.5875	1.0125	1.2125
12	1.3500	0.7500	0.9375
13	1.1375	0.5250	0.6875
14	0.9500	0.3250	0.5875
15	0.7750	0.1500	0.3625
16	0.6375	0	0.1375

Figure 5.27: Focal law values of each element at receiving (blue) for geometrical compensation of Notch B by the compensation of the base material where the sound waves pass through the weld region (green) and (2) for the compensation of the time delay laws (red)

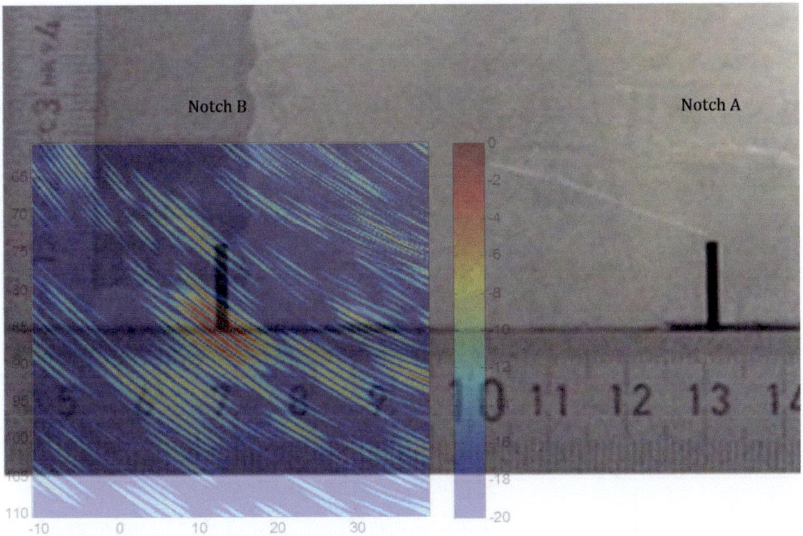

Figure 5.28: Image gained by TFM for Notch B after both the compensation of the base material where the sound waves pass through the weld region by the geometrical computation and the geometrical computation based compensation of the misplacement.

5.6 Results

This chapter shows improvements of ultrasonic signals applying compensation procedures in this thesis. The delay laws of Notch A and B were determined by RDLT. When the sound wave velocity of the material is accurate, the position of the beam focusing will be accurate as well. For the reliable testing of a dissimilar metal weld, accuracy of the calculations is mainly depending on a function of the wave velocity. Hence, only the delay laws determined are inserted in the software of the PcusPro and the testing measurements are performed by the same parameters at the same scan positions.

Figure 5.29 shows the S-scan images focused at root-edge of Notch A before and after the compensation of the delay laws. Both images are the result of the sound waves which are reflected at the edge of the notch by "angle mirror effect". The image gained before the compensation indicates a 5.5mm misplacement. This misplacement can be identified by both, incident angle (24.7°) and the wave propagation distance (125.6mm), which are acquired from the coordinate profile according to Figure 5.6. A square frame in black line containing the signals reflected from the edge is enlarged before and after the compensation for better identification of the signal quality. The signals after compensation in spectral scale show the higher SNR with the enhanced amplitude in comparison with the signals before compensation. The angle steered to focus the root-edge became smaller (22.6°) and the beam path was shortened (122.5mm). Finally it was noticed that the signals are more precisely located.

Figure 5.30 shows the signals reflected from the tip of Notch A before and after the compensation. The both images are the result of the sound waves which are focused at the notch tip. The misplacement can be identified by both the incident angle (28°) and the wave propagation distance (119.5mm). A square frame in black line containing the signals reflected from the edge is enlarged before and after the compensation for better identification of the signal quality. The results also reveal the significantly improved SNR with the higher lateral resolution. Both the steering angle (25.6°) and the beam path (115.3mm) to the target were adjusted to focus the target. As a result, the tip signal indicates more precisely in original position.

Figure 5.31 highlights the result of this approach. Both images are obtained by the sound waves which are reflected from the edge of the notch by "angle mirror effect". The image gained before the compensation shows signal misplacement in two axes, i.e., 5.5mm in horizontal and 5.1mm in vertical axis.

The misplacement can be identified with both the incident angle (20.6°) and the wave propagation distance (138.2mm). At the image before and after the compensation a square frame in black line containing the signals is enlarged to help identifying the signal quality. The S-scan image after compensation shows more accurately located position. Even though the SNR was not highly increased, the image allowed the improved POD. Unlike the case of Notch A, the steered angle to the target became smaller (17.2°) but the beam path was 0.5 mm longer (138.7mm) after the compensation. With the compensation, the signals reflected from the bottom of the weld root area became closer in the original position. Quantifying the Notch B image was not accomplished because the notch-tip signal was not found in S-scan image although FMC-TFM provided both tip and rood-edge signals by reconstruction.

Figure 5.32 represents the monitored signals reflected inner weld. This signal can be considered as scattered mainly due to an internal sound reflector. Before the reconstruction, the signals are identified with the incident angle (21.1°) and the wave propagation distance (73.5mm). After the reconstruction by RDLT, the coordinate of the signal is also moved along with the signal reflected from the root-edge of Notch B representing the incident angle (17.7°) and the wave propagation distance (74mm). It is interesting to notice that the signals after compensation of Notch B became enhanced with increased SNR as well. However, the movement in the coordinate would not mean the compensation of the misplacement.

Figure 5.33 represents the whole image including the coordinate data of the signal location after the compensation.

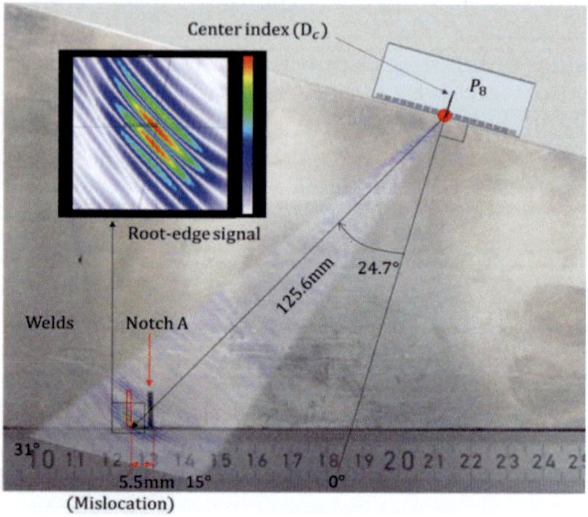

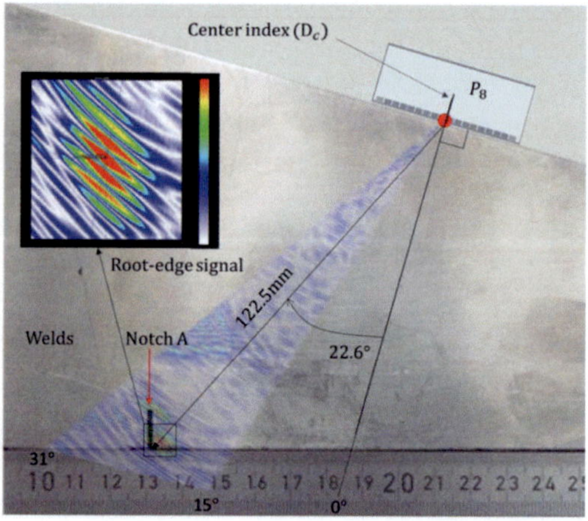

Figure 5.29: S-scan images scanned at Notch A focusing across root-edge of notch representing misplacement before compensation (top, the same image as Figure 5.8) and original position after compensation (bottom). The top image shows a 5.5mm misplacement which is identified with both the incident angle (24.7°) and the wave propagation distance (125.6mm). A square frame in black line containing the signals reflected at the edge is enlarged for better identification of the signal quality. After the compensation, the location of the signals are identified with the incident angle (22.6°) and the wave propagation distance (122.5mm).

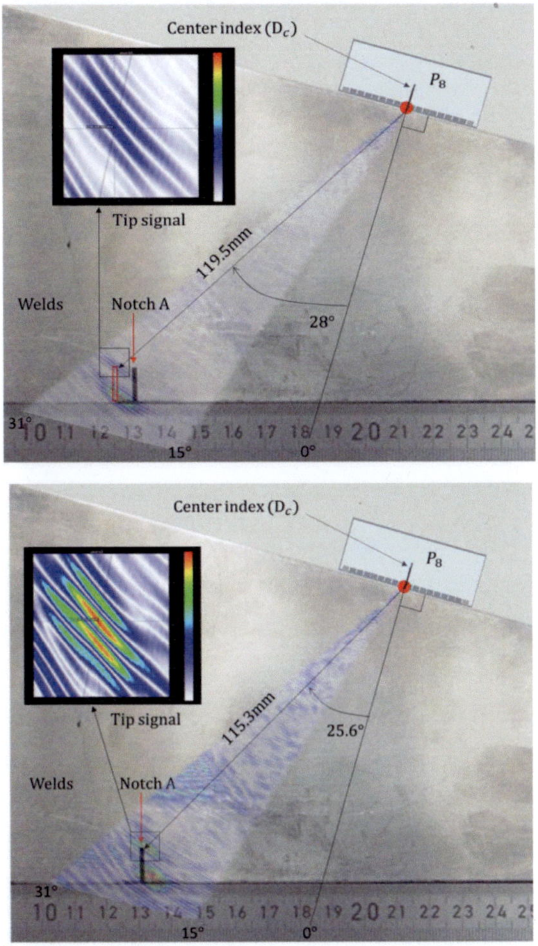

Figure 5.30: S-scan images scanned across Notch A focusing at notch-tip representing misplacement before compensation (top, the same image as Figure 5.8) and original position after compensation (bottom). The top image shows a misplacement of the incident angle (28°) and the wave propagation distance (119.5mm). A square frame in black line containing the signals reflected at the edge is enlarged for better identification of the signal quality. After the compensation, the location of the signals are identified with the incident angle (25.6°) and the wave propagation distance (115.3mm).

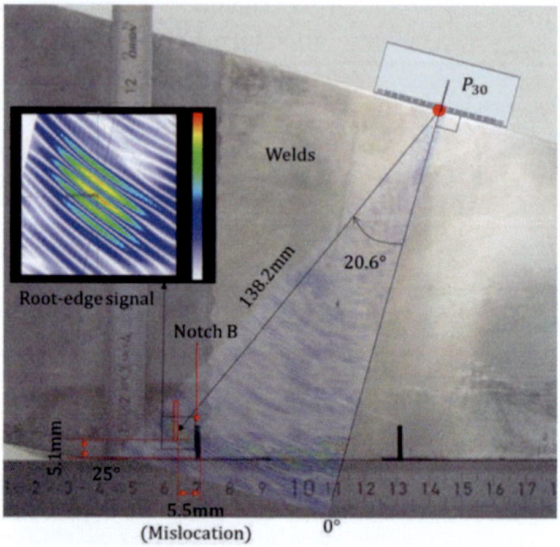

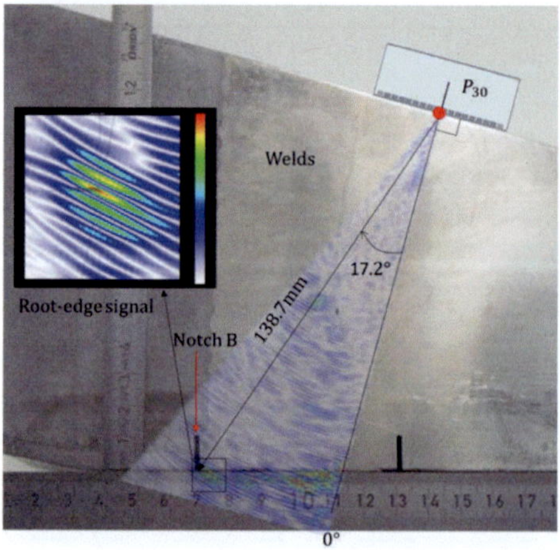

Figure 5.31: S-scan images scanned across Notch B focusing at root-edge representing misplacement before compensation (top, the same image as Figure 5.9) and original position after compensation (bottom). The misplacement is represented in two axes, i.e., 5.5mm in horizontal and 5.1mm in vertical axis which is identified with both the incident angle (20.6°) and the wave propagation distance (138.2mm). A square frame in black line containing the signals is enlarged for better identification of the signal quality. After the compensation, the location of the signals are identified with the incident angle (17.2°) and the wave propagation distance (138.7mm).

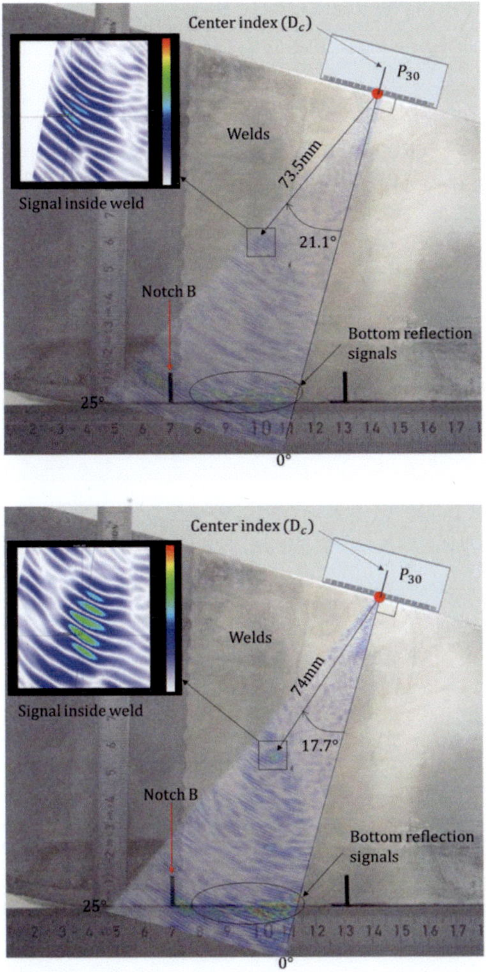

Figure 5.32: S-scan images representing signals inner weld unfocused before RDLT of Notch B (top, the same image as Figure 5.8) and after RDLT (bottom). Before the reconstruction the signals considered as scatter are identified with the incident angle (21.1°) and the wave propagation distance (73.5mm). After the reconstruction, the coordinate of the signal is also moved along with the signal reflected at the root-edge of Notch B indicating the incident angle (17.7°) and the wave propagation distance (74mm).

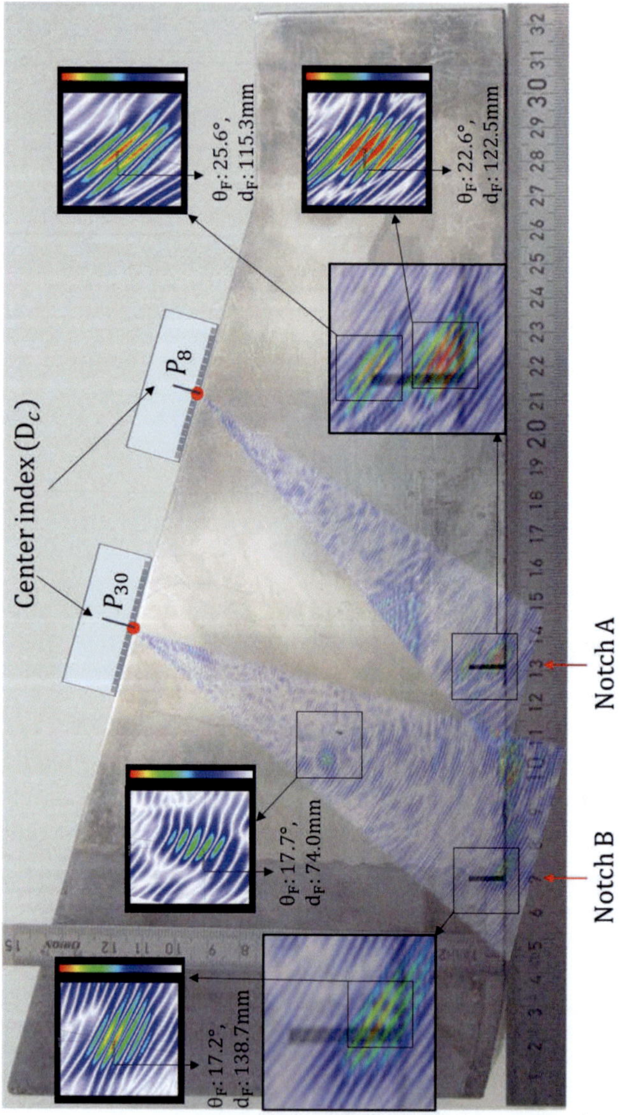

Figure 5.33: Sectorial scan images scanned across Notch A and B in original position corrected by calculated time delay laws

Chapter 6. Conclusion

6.1 General summary

The objective of this thesis has been to investigate the wave propagation through the austenitic dissimilar metal weld (DMW) with the definite aim of the improvement of the probability of detection (POD) and the reduction of the individual skill difference of sizing flaws. The approach is to understand the key principle of the referencing delay law technique (RDLT), to apply it to the current existing ultrasonic testing method and then to aid the quantification of the flaws more precisely which was motivated from the challenges due to the characteristics of the anisotropic and inhomogeneous media, e.g. the complex steering of the wave paths in the material.

As the advances in computing power and affordability have been significantly developed in recent years, the corresponding technologies and new methods, for example full matrix capture (FMC) and total focusing method (TFM) have been introduced and upgraded.

The profile of the specimen was provided in Chapter 3. In order to understand the complicated grain structures of the weld thoroughly, grazing incidence ultrasound microscopy (GIUM), etching metallography and electron backscatter diffraction (EBSD) were performed. In addition, scanning laser vibrometry (SLV) was fulfilled for the visualization of the ultrasonic paths in the weld. By the visualized images noticeable wave paths in the anisotropic media were illustrated: the distinction of the wave front between each boundary which is useful for explaining the anisotropic phenomena encountered in the thesis. The contrast with the wave propagation under the condition of isotropic and homogeneous material was studied by numerical simulation using FE modeling and the corresponding properties of the wave were summarized. Two artificial notches were then produced by means of electronic discharged machining. One notch is placed in the transversely isotropic region and the other one is behind the weld. In order to demonstrate the evidence of any defect inside the specimen, the immersion tank ultrasonic testing was performed for the whole volume of the specimen.

In Chapter 4 the principle of both the data acquisition by FMC and the processing by TFM and SAFT was described. The chapter introduced the new technique to solve the challenges by RDLT and described the principle. In that section the procedure how to reference the data and to correct the error mainly due to the difference of the sound velocity and the skew was described. The chapter concluded with a preliminary feasibility study on RDLT by using the austenitic calibration block. The study provided the possibilities both of measuring the direct sound velocities at a certain scan point which is used to determine the focal laws and of enhancing signal-to-noise ratio (SNR) with locating the notch-tip. In particular it represented the higher POD by locating the minor signal reflected from the side drilled hole in 1.5mm diameter of which the size is less than half of the wave length of the testing transducer.

The measurements were described in Chapter 5. The measurements were achieved both by the phased array ultrasonic testing and by FMC. The S-scan images gained by the phased array testing represented the misplacement and the missizing of the signals acquired from Notch A and B and provided the information of the degradation in coordinates with the focal depth and the steered angle to the reflector. The image processed by TFM also showed the degradation in horizontal and in vertical axes. This chapter described how RDLT compensates the misplaced and missized image by means of the geometric based computation. The results from two notches served to highlight the importance of utilizing the time delay laws that represent the anisotropic and inhomogeneous properties of the weld to overcome the misplacement and missizing of the defect. The determined time delays were used to correct the degraded images. Through this compensation process, the misplaced and missized images were reshaped because the original geometry in which the waves travel with varying velocity was changed into the geometry in which the sound propagated with constant velocity. The conventional phased array testing by S-scan image was employed for validation purposes and the images showed the more precisely located signals with improved SNR.

6.2 Review of findings

6.2.1 Grain structures of anisotropic weld

By three different analysis methods, e.g. GIUM, etching metallography and EBSD, the grain structures of the material anisotropy were investigated. The results revealed the strong dendritic patterns of the grain with the vertical direction corresponding to the observed solidification growth of the weld. It was found that the major grain size is 2 to 4 mm and tilted from the vertical axis by up to 45° which represents strong anisotropy and heterogeneity. The grain orientations in the buttering region are directed mostly in horizontal axis.

6.2.2 Visualized wave propagation in anisotropic weld

The visualization of the P-wave path was fulfilled by using SLV. The results show:

i. the wave propagation in the anisotropic media: the deformed wave at planar interface between the transversely isotropic and the anisotropic and the squashed wave passing through the anisotropic media

ii. the undeformed wave passing out of the anisotropic media.

6.2.3 Simulation of wave propagation in isotropic media

Finite element simulation of the wave propagation in isotropic and homogeneous media was performed as if the specimen was the elastically constant. This approach helps to understand the phased array principle and its application to imaging in isotropic media like mild steel material.

6.2.4 Reliable sound velocity measurement by RDLT

The core part of this thesis was to achieve how reliably the sound wave velocity can be directly measured within the interested scanning range by single scan point in the specimen. Thanks to the powerful data set and processing technique by FMC-TFM-SAFT, the velocity values over the range in horizontal and in vertical directions are sufficiently stable. The results of this investigation are listed in the Appendix.

6.2.5 Scattered signals at interface between transversely isotropic and anisotropic media

With the aid of TFM processing, some scattered signals at the boundary between the transversely isotropic and the lower part of the anisotropic weld region were found. The signals can be useful for fitting the image of TFM to the image of the specimen.

6.2.6 Geometric based computation for time delay

The key part of this thesis was to use the pre-known geometric data for calculating time delay with the aid of the sound velocity values acquired. The height of the notch became the reference to correct the missizing error and using the distance of two notches the misplacement can be corrected by means of the referencing principle. The results improved sizing and locating of the notches.

6.2.7 Notch tip signal behind the austenitic weld

By using FMC-TFM, the notch tip signal behind the austenitic weld was clearly found. Although S-scan images cannot show the tip signal after the compensation of the time delay, the reconstructed images by FMC-TFM provide the tip signal. After correcting the position by RDLT, the corresponded data can be used for sizing the depth or height of the flaw. This can reduce the individual skill deviation of NDT inspectors.

6.2.8 Improvements of SNR and POD

Through the validation by the phased array ultrasonic testing with the corrected focal laws, the S-scan images of Notch A and B showed significantly improved SNR and increased POD correspondingly. The results also validate the enhanced quantification of the sizing and the locating of the defects.

6.3 Areas for improvements

In this section some limitations are discussed. RDLT has taken the basic role of quantifying the flaw signals. For this reason, a reference block is necessary and sound reflectors for referencing are required. When the position of the testing defect is specified as it is often organized in the nuclear power plant industry like the performance demonstration, RDLT can be a promising method for it. However, when a defect should be detected in various positions in the test piece, it requires the corresponding numbers of artificial sound reflectors which causes the time and the cost to be increased.

For application of FMC and TFM, since it uses a single element for transmitting the pulse, the total output of the acoustic power is accordingly limited which can be restricted for travelling through the huge size testing specimen in depth.

In addition, a high-speed computing process equipped system is necessary to calculate and to process the data.

6.4 Perspective

The ability to parameterize the geometry and the directly measured sound velocity is the key to RDLT. It is conceivable that increase in the number of the array elements and the acoustic power would allow RDLT to increase SNR. Doing so would allow the user of RDLT to detect challenging flaws, for example stress corrosion fatigue cracks in the weld.

As a future work in perspective view point, the following investigations are suggested of which the realization can become a testing field-oriented solution:

i. developing 3D FMC-TFM based RDLT with improved SNR processing technique for optimized signal interpretation of the defects,

ii. enhancing the total output of the acoustic power generated at a single element for increasing the effective penetration depth,

iii. application of RDLT for field testing for detection of real discontinuities in the anisotropic and heterogeneous weld

i) it is necessary to produce a reference block with matrix-like side-drilled holes which can provide wave velocity values in multi direction

ii) the sound velocities measured can be used as data base for focal law calculation and be set to a phased array ultrasonic testing for a testing specimen in the field.

Appendix

A. 1 Chemical composition of filling metal ERNiCrFe-7 for weld

Element	Ni	C	Mn	Fe	S	Cu	Si
Content (%)	Bal.	Max 0.04	Max 1.0	7.0-11.0	Max 0.015	Max 0.30	Max 0.50
Element	Al	Cr	Ti	Nb+Ta	Mo	P	-
Content (%)	Max 1.10	28.0-31.5	Max 1.0	Max 0.10	Max 0.50	Max 0.02	-

(by the courtesy of Korea Hydro and Nuclear Power R&D Institute)

A. 2 Data for modeling of transducer and for numerical simulation

A.2.1. Transducer, discretization and material parameters

Transducer properties	
Central of frequency	1.5 MHz
Transducer element length	2.8 mm
Number of elements	16
Discretization parameters	
Time step size	2e-8
Element size	0.2 mm ($\boldsymbol{\lambda/20}$)
Total number of nodes	838434
Total number of elements	837533
Element type	Plane strain element (Elform13)
Material properties	
Young's modulus	210Gpa
Density	7850 Kg/m^3
Poisson ratio's	0.3

A.2.2. Time delay values calculated for Notch A and Notch B

Element	Time delay[μs] Notch A	Time delay[μs] Notch B
1	0.0000	0.0000
2	0.2127	0.1903
3	0.4253	0.3807
4	0.6380	0.5710
5	0.8507	0.7613
6	1.0633	0.9517
7	1.2760	1.1420
8	1.4886	1.3324
9	1.7013	1.5227
10	1.9140	1.7130
11	2.1266	1.9034
12	2.3393	2.0937
13	2.5520	2.2840
14	2.7646	2.4744
15	2.9773	2.6647
16	3.1899	2.8551

A. 3 Sound velocity values of calibration block in 0 degree angle by using RDLT

Calibration Block (K1)			
Scanning point	Velocity (m/sec)	Scanning point	Velocity (m/sec)
16	5,751.3	32	5,749.2
17	5,749.2	33	5,755.4
18	5,749.2	34	5,749.2
19	5,745.1	35	5,740.9
20	5,747.1	36	5,743.0
21	5,751.3	37	5,747.1
22	5,747.1	38	5,743.0
23	5,749.2	39	5,745.1
24	5,749.2	40	5,745.1
25	5,751.3	41	5,745.1
26	5,753.3	42	5,747.1
27	5,753.3	43	5,747.1
28	5,753.3	44	5,753.3
29	5,753.3	45	5,740.9
30	5,749.2	46	5,755.4
31	5,747.1		
Mean velocity	5,748.6	Standard deviation	4.0

A. 4 Sound velocity values of reference block in 0 degree angle by using RDLT

Scan point	Time (μs)	Depth (d_{xn}, mm)	Velocity (m/s)	Scan point	Time (μs)	Depth (d_{xn}, mm)	Velocity (m/s)
1	28.0	78.6	5,609.6	19	33.0	92.5	5,603.8
2	28.3	79.4	5,614.6	20	33.3	93.3	5,605.9
3	28.6	80.2	5,614.6	21	33.6	94.1	5,606.0
4	28.9	81.0	5,612.2	22	33.8	94.8	5,608.1
5	29.1	81.7	5,614.6	23	34.1	95.6	5,610.2
6	29.4	82.5	5,614.6	24	34.4	96.4	5,612.3
7	29.7	83.3	5,614.5	25	34.6	97.2	5,614.3
8	29.9	84.0	5,614.5	26	34.9	97.9	5,614.3
9	30.2	84.8	5,614.5	27	35.2	98.7	5,614.3
10	30.5	85.6	5,612.2	28	35.4	99.5	5,614.3
11	30.8	86.4	5,614.5	29	35.7	100.2	5,614.3
12	31.0	87.1	5,614.5	30	36.0	101.0	5,614.3
13	31.3	87.9	5,616.7	31	36.2	101.8	5,618.1
14	31.6	88.7	5,616.7	32	36.5	102.6	5,618.1
15	31.9	89.4	5,614.4	33	36.8	103.3	5,610.4
16	32.2	90.2	5,610.1	34	37.1	104.1	5,610.4
17	32.4	91.0	5,610.1	35	37.4	104.9	5,614.2
18	32.7	91.8	5,608.0	Average velocity			5,612.6
Standard deviation		Time difference (μs)		Linear fit	Interval distance, x (mm)		
3.4		0.3		22.002	0.8		

A. 5 Sound velocity values of reference block in 90 degree angle by using RDLT

Scan point (P_n)	Velocity (m/s)
16	5,710.7
17	5,727.7
18	5,727.7
19	5,711.2
20	5,712.1
21	5,702.0
22	5,702.0
23	5,702.0
24	5,698.3
25	5,725.9
26	5,725.9
27	5,722.2
28	5,707.0
Standard deviation	11.1
Average velocity	5,713.4

References

[1] G Van Goethem et al., " EU research in "operational safety of existing installations" under the nuclear fission programme 1998–2002", Nuclear engineering and desing, Vol.221, pp.3-22 (2003)

[2] G. J. Davies and J.G. Garland, "Solidification structures and properties of fusion welds", International metallurgical reviews, Vol.20 (1975)

[3] HG. Lambers, S. Tschumak, HJ. Maier and D. Canadinc, "Role of austenitization and pre-deformation on the kinetics of the isothermal bainitic transformation", Metal Mater Trans A. 40 (6) (2009).

[4] X. Edelmann, "Ultrasonic examination of austenitic welds at reactor pressure vessels", Nuclear Eng. & Design 129: pp.341-355 (1991).

[5] J. Lenkkeri, A. Juva, "The effect of anisotropy on the propagation of ultrasonic waves in austenitic stainless steel", Reliability of ultrasonic inspection of austenitic materials, Belgium (1980).

[6] X. Edelmann, „Application of ultrasonic testing techniques on austenitic welds for fabrication and in-service inspection", NDT International, Vol.14, pp.125-133 (1981).

[7] H. Gao, S. M. Ali and B. Lopez, "Inspection of austenitic weld with EMATS", AIP Conf. Proc. 1211, pp.1175-1181 (2009).

[8] M. Kroening, A. Bulavinov and F. Walte, "Improving the inspectability of stainless steel and dissimilar metal welded joints using inverse phase-matching of phased array time-domain signals", 17th WCNDT, Shanghai, China (2008)

[9] C. nageswaran, C. Carpentier and Y. Y. Tse, "Microstructural quantification, modeling and array ultrasonics to improve the inspection of austenitic welds", Insight- Nondestructive testing and condition monitoring, Vol.51, No.12, pp.660-666 (2009).

[10] F. Ammirato and R. Shankar, "Ultrasonic examination of dissimilar metal welds in BWR and PWR plants", International Jr. of Pressure Vessels and Piping, Vol.35, No.11, pp.37-56 (1988).

[11] J. Ye, H. Kim and S. Song, "Model-based simulation of focused beam fields produced by a phased array ultrasonic transducer in dissimilar metal welds", NDT & E International, Vol.44, No.3, pp.290-296 (2011).

[12] X. Edelmann, "Ultrasonic examination of austenitic welds at reactor pressure vessels", Nuclear Engineering and Design, Vol.129, No.3, pp.341-355 (1991).

[13] M. Beth and J. L. Monjaret, "Round robin test results on ultrasonic testing of austenitic welds," 2nd international conference on NDE in relation to structural integrity for nuclear and pressurized components, New Orleans, Vol.6, No.1, (2001).

[14] S. Kang, Y. Choi, "Status of performance demonstration for in-service NDE of nuclear power plant," Summer Conference of KSNT, pp.115-123 (1999).

[15] ASME B & PV Code Section XI, "Rule for in-service inspection of nuclear power plant components," edition & addenda (1995).

[16] B. Yoon and et. al, "Development of an ultrasonic performance demonstration system for dissimilar metal welds in Korean nuclear power plants," Jr. of Nuclear Science and Technology, Vol.48, No.11, pp.1403-1411 (2011).

[17] B. Yoon and et. al, "Round robin test for performance demonstration system of ultrasonic examination personnel," Jr. of KSNT, Vol. 24, No. 4, pp. 378-383 (2004).

[18] O. T. Von Ramm and S. W. Smith, "Beam steering with linear arrays", IEEE Transactions on biomedical engineering, Vol.BME-30, No. 8 (1983).

[19] C. Clay, S. Wooh, L. Azar and J. Wang, "Experimental study of phased array beam steering characteristics", Jr. of nondestructive evaluation, Vol.18, No.2 (1999).

[20] J. A. Ogilvy, "Model for predicting ultrasonic pulse-echo probability of detection", NDT & E international, Vol.26, No.1, pp.19-29 (1993).

[21] J. Saniie and N. M. Bilgutay, "Quantitative grain size evaluation using backscattered echoes", Jr. of the acoustical society of America, Vol.80, No.6, pp.1816-1824 (1986).

[22] P. Emmanuel, "Ultrasonic attenuation and velocity in three transformation products in steel", Jr. of applied physics, Vol.35, No.5, pp.1474-1482 (1964).

[23] F. Schubert, "An ultrasonic mode conversion technique for characterizing prism-shaped material samples - experimental and numerical results", ECNDT (2006).

[24] Olympus NDT, "Advances in phased array ultrasonic technology applications", ISBN: 0-9735933-4-2 (2007).

[25] H. Cho, "Development of image processing techniques for ultrasonic flaw sizing", Dissertation, Seungkyungwan University (2007).

[26] A. W. Elbern, L. Guimarães, "Synthetic aperture focusing technique for image restoration", International symposium on NDT contribution to the infrastructure safety systems (1999).

[27] C. Holmes, B. Drinkwater, and P. Wilcox, "Post-processing of the full matrix of ultrasonic transmit–receive array data for non-destructive evaluation", NDT & E International, vol. 38, no. 8, pp.701-711, Dec. (2005).

[28] SW. Smith, GE. Trahey, OT. von Romm, "Two-dimensional arrays for medical ultrasound", Ultrasonic Imaging 14, pp.213–233 (1992).

[29] S.-C.Wooh and Y. Shi, "Influence of phased array element size on beam steering behavior", Ultrasonics, vol.36, pp.737-749 (1998)

[30] R/D Tech Corp, "Phased Array Technical Guidelines: Useful Formulas, Graphs, and Examples", Waltham, MA: R/D Tech Corp.,(2005).

[31] K. Nakahata and N. Kono, "3-D modelings of an ultrasonic phased array transducer and its radiation properties in solid", Ultrasonic Waves, Dr Santos (Ed.), ISBN: 978-953-51-0201-4, pp.62-63 (2012).

[32] Th. L. Szabo, "Diagnostic ultrasound imaging", ISBN: 0-12-680145-2, pp.188-189 (2004).

[33] S. C. Wooh, and Y. Shi, "Optimization of ultrasonic phased arrays", Review of Progress in Quantitative Nondestructive Evaluation, pp.883-890 (1998).

[34] S. R. Doctor, "Nuclear power plant NDE challenges - past, present, and future", AIP Conf. Proc. 894, pp.17-31 (2006).

[35] R. B. Bhavsar and E. L. Hibner, "Evaluation of testing techniques for selection of corrosion resistant alloys for sour gas service", corrosion/96, paper no. 59, NACE International, Houston, TX, USA, (1996).

[36] J. Takatsubo, B. Wang, H. Miyauchi, H. Tsuda, N. Toyama, K. Urabe, "Laser ultrasonic technique for visualizing ultrasonic waves propagating on 3-D object", 17th WCNDT, Shanghai, China (2008).

[37] F. Schubert, H. Scholz, M. Froehlich, M. Barth, C. Richter, Z. Bor, S. Heilmann, P. Heilmann, N. Meyendorf, "A New Model-Based Approach for Ultrasonic Testing of Dissimilar Welds", 6th NDT in Progress (2011).

[38] M. Barth, B. Köhler, L. Schubert, "3D-Visualization of Lamb waves by laser vibrometry", Proceedings of the Fouth European Workshop Structural Health Monitoring, DES tech Publications, Inc. Lancaster, pp. 641-648 (2008).

[39] B. Köhler, M. Barth, P. Krueger, and F. Schubert, "Grain structure visualization with surface skimming ultrasonic waves detected by laser vibrometry", Appl. Phys. Lett. 101, 074101 (2012).

[40] S. Crampin, "An introduction to wave propagation in anisotropic media", Geophys. J. R. astr. Soc.76, pp. 17-28 (1984).

[41] M. Barth, B. Köhler, F. Schubert, "Grain structure visualization of austenitic Welds by laser detection of grazing incident ultrasonic wave fields", NDCM-XIII, Le Mans, (2013).

[42] S. Piazolo, V. G. Sursaeva, and D. J. Prior, Mater. Sci.Forum 495-497, 213 (2005).

[43] V. Randle and O. Engler, "Introduction to texture analysis: Macrotexture, microtexture and orientation mapping", pp.125-203. CRC Press, New York (2000).

[44] R. A. Schwarzer, D. P. Field, B. L. Adams, M. Kumar, and A. J. Schwartz, "Electron backscatter diffraction in materials science", 2nd ed., (eds. A. J.

Schwartz, M. Kumar, B. L. Adams, D. P. Field), pp.1-20, Springer, New York (2009)

[45] F. J. Humphreys, "Proc. ICOTOM 12 (ed. J. Szpunar)", pp.74-79, NRC Research Press, Ottawa (1999)

[46] F. J. Humphreys, "Characterisation of fine-scale microstructures by electron backscatter diffraction (EBSD)", Scripta Materialia, Vol.51, No.8, pp.771-776 (2004)

[47] Randle V. and Engler O., "Introduction to Texture Analysis", CRC Press, (2000).

[48] A. Bóna, and M.A. Slawinski, "Fermat's principle for seismic rays in elastic media," Jr. Appl. Geophysics Vol.54, pp. 445-451 (2003).

[49] R. Ludwig. and W. Lord, "A finite element formulation for the study of ultrasonic NDT phenomena.", IEEE Transactions on Ultrasonics, Ferroelectrics and Frequency Control, Vol.35, pp.809-820 (1998).

[50] B.Ganesan, V.Kommareddy, S.Ramaswamy and K.Manoj, "Application of explicit finite element method(LS-DYNA) to simulate ultrasonic wave propagation", Altair CAE User Conference, (2005).

[51] J.Krautkramer and H Krautkrammer, "Ultrasonic testing of materials", Spring-Verlag, Fourth edition, (1990).

[52] Karl J. Langenberg, Ren´e Marklein , "Transient elastic waves applied to nondestructive testing of transversely isotropic lossless materials: a coordinate-free approach", Wave Motion Vol.41, pp.247–261 (2005).

[53] H. Yim, "Analysis of ultrasonic wave propagation in transversely isotropic austenitic welds", Journal of KSNT/SC0007 (2001).

[54] C. Holmes, B. Drinkwater, and P. Wilcox, "Post-processing of the full matrix of ultrasonic transmit-receive array data for non-destructive evaluation", NDT & E International, vol.38, No.8, pp.701-711 (2005).

[55] J. A. Jensen, S. I. Nikolov, K. L. Gammelmark, and M. H. Pedersen, "Synthetic aperture ultrasound imaging." Ultrasonics, vol.44 (2006).

[56] K. Thomenius, "Evolution of ultrasound beamformers", IEEE Ultrasonics Symposium, Vol.2, pp.1615-1622 (1996).

[57] M. Weston, "Development of ultrasonic full matrix capture technology to improve flaw characterization", TWI Members Report 966 (2010).

[58] C. Holmes, B. Drinkwater, and P. Wilcox, "Exploiting the full data set from ultrasonic arrays by post-processing", Review of progress in quantitative nondestructive evaluation, vol.25, pp.845-852 (2006).

[59] A. McNab, I. Stumpf, "Monolithic phased array for the transmission of ultrasound in NDT", Ultrasonics, Vol.24, No.3, pp.148-155 (1986)

[60] P.D. Wilcox, C. Holmes and B.W. Drinkwater, "Advanced reflector characterization with ultrasonic phased arrays in NDE applications", IEEE Transactions on Ultrasonics, Vol.54, No.8, pp.1541-1550 (2007).

[61] T. K. Song, S. B. Park, "A new digital phased array system for dynamic focusing and steering with reduced sampling rate", Ultrasonic Imaging, Vol.12, No.1, pp.1-16 (1990).

[62] R. Long et al., "Ultrasonic phased array inspection of flaws on weld fusion faces using full matrix capture", Review of Progress in Quantitative Nondestructive Evaluation, pp.848-855 (2008).

[63] R Long, J Russel and P Cawley, "Through-weld ultrasonic phased array inspection using full matrix capture", In Review of Progress in Quantitative

Nondestructive Evaluation (QNDE), Vol.29, AIP conference proceedings, Vol.1211, pp.918-925 (2010).

[64] G.D. Connolly, "Modelling of the propagation of ultrasound through austenitic steel welds", Dissertation, Imperial College London, pp.126-128 (2009).

[65] M. Spies, H. Rieder, "Synthetic aperture focusing of ultrasonic inspection data to enhance the probability of detection in strongly attenuating materials", NDT E Int. Vol.43, pp.425–431 (2010).

[66] T. Stepinski, "Synthetic aperture focusing technique in ultrasonic inspection of coarse grained material", SKI Report, ISSN 1104-1374, ISRN SKI-R-08/06-SE (2007).

[67] A.J. Hunter, B.W. Drinkwater, and P.D. Wilcox, "The wavenumber algorithm for full-matrix imaging using an ultrasonic array", IEEE Trans. on Ultrasonics, Ferroelectrics & Frequency Control, Vol.55, No.11, pp.2450-2462 (2008).

[68] J. Bilitz, G. Simpson, "Ultrasonic Methods of Nondestructive Testing", Springer, ISBN 0412604701, pp. 167-175 (1996).

[69] ASTM E 428-05a, "Standard practice for fabrication and control of metal, other than aluminum reference, blocks used in ultrasonic examination".

[70] R. A. Roberts, "Ultrasonic beam transmission at the interface between an isotropic and a transversely isotropic solid half-space," Jr. of Ultrasonics, Vol.26, pp.139-147 (1988).

[71] K. Nakahata, S. Hirose, F. Schubert, B. Koehler, M. Barth "Image based simulation of wave propagation in dissimilar metal welding part and its verification," Jr. of JNDT, Vol.10, No.2 (2011).

[72] R. H. Burkel, D. J. Sturges, W. t. Tucker and R. S. Gilmore, "Probability of detection for applied ultrasonic inspections", Review of progress in QNDE, 15B, D.O. Thompson and D.E. Chimenti, Eds., pp.1991-1998 (1996).

[73] Chapter 5: F. Lewis et al, "Ultrasound transducer models for piezoelectric polymer films", IEEE Transactions on ultrasonics. Ferroelectrics, and frequency control, Vol.*36*. No.3, (1989).

[74] M. Willatzcn, "Ultrasound transducer modeling-general theory and applications to ultrasound reciprocal systems", IEEE Transactions on ultrasonics, Ferroelectrics, and frequency control, Vol.48, No.1, (2001).

[75] L. Schmerr, "Fundamentals of ultrasonic nondestructive evaluation: A modeling approach", Springer, ISBN 0306457520 (1998).

[76] J.A. Ogilvy, "Ultrasonic beam profiles and beam propagation in an austenitic weld using a theoretical ray tracing model", Ultrasonics, Vol.24, No.6, pp.337 -347 (1986).

[77] K.J. Langenberg et al, "Application of modeling techniques for ultrasonic austenitic weld inspection", NDT&E International , Vol.33, pp.465–480 (2000).

[78] M. Spies, "Modeling transient radiation of ultrasonic transducers in anisotropic materials including wave attenuation", AIP Conf. Proc., Vol.615, No.807, doi: 10.1063/1.1472880 (2002).

[79] R. B. Yang, A. K. Mal, "Multiple scattering of elastic waves in a fiber-reinforced composite", Jr. Mech. Phys. Solids, Vol.42, No.12, pp.1945-1968 (1994).

[80] A. B. Bouda, A. Benchaala, S. Lebaili, "Grain size influence on ultrasonic velocities and attenuation", NDT & E International, Vol.36, No.1, pp.1-5 (2003).

[81] PCUS® PRO ARRAY, http://www.izfp-d.fraunhofer.de/assets/downloads/PCUS_pro_Array.pdf

List of symbols

A	wave amplitude, polarisation amplitude
a	width of single element of phased array
C	stiffness matrix of elastic constants
C	correlation coefficient
c, c^2	sound velocity
D	active aperture of phased array
D	depth
d	pitch width
d	distance
F	focal depth/point
F( )	Fourier transform of
f	frequency
G( )	Complex spectrum of
I	image
I	image response matrix
K	function of velocity difference

L	elevation of phased array element
M	relevant modulus
N	the number of phased array elements
n	the number of (subscripts)
p	pressure
P	scan position
p	directivity function
R	reflection coefficient
S	signal data matrix
s	step
s	sampling
T	wave period
t	period of beam propagation
u	received signals from the reference block
$\tilde{u}$	ideal signals based on geometric computation of reference block
z	acoustic impedance
α	angle in medium 1
β	angle in medium 2
λ	wavelength
ρ	density
τ	time of flight, Fermat time
θ	wave angle in x-z plane
ϕ	wave angle in x-y plane
ω	angular frequency

Superscripts

cr	critical
eff	effective
F	flaw/ discontinuity
I	incident
I	ideal
i	initial
ij	directionality of elastic constants
ij	sequence of element number for transmitting and receiving
min	minimum
mod	modulation
P	longitudinal
R	reference block/ specimen
R	reconstruction
r	reflection/reflected
S	transverse
s	steering angle

List of abbreviations

BLM	beam locus mapping
CRA	corrosion resistant alloy
CSM	common source method
DMW	dissimilar metal welds
EBSD	electron backscatter diffraction
EDM	electrical discharge machining
EFIT	elastodynamic finite integration technique
FCC	face-centred cubic
FFT	fast fourier transform
FMC	full matrix capture
GIUM	grazing incidence ultrasound microscopy
GTAW	gas tungsten arc weld
HAZ	heat affected zone
IGSCC	inter-granular stress corrosion cracking
NDT	nod-destructive testing
PD	performance demonstration

PISC	program for the inspection of steel components
POD	probability of detection
RDLT	referencing delay law technique
RMSE	root mean square error
ROI	region of interest
SAFT	synthetic aperture focusing technique
SEM	scanning electron microscope
SLV	scanning laser vibrometry
SMAW	shielded metal arc weld
SNR	signal-to-noise ratio
TFM	total focusing method
TOF	time of flight

List of figures

List of tables

Acknowledgements

I would like to express the deepest appreciation to my advising Prof. Norbert Meyendorf who kept supporting not only my research but also life in Germany. Without his guidance this study would not have been possible. I also would like to express thanks to Dr. Frank Schubert who gave me direction when it was most needed. His advice for my study kept me focused on this topic. I especially thanks to Prof. Jongbum Seo whose encouragement, supervision and support from the preliminary to the concluding level enabled me to develop an understanding of the subject. I deeply appreciate for Prof. Juergen Schreiber who provided support and instilled mental stimulation in regards to my research. I am also heartily thankful to Prof. Ohyang Kwon, Prof. Hansung Kim and Dr. Joerg Opitz who inspired me to take pride in my research. Without their persistence, I would have lost my motivation. I also offer my best regards and blessings to Mr. Seungpyo Lee and Dr. Byungsik Yoon who supported me for the specimens which is difficult to obtain.

Also, I am grateful to my friends Kyounghee, Kyungho, Kyoungjoo, Martin and Andreas, who helped me tremendously. I also thank my mother and mother in law and father and father in law who prayed for me a lot.

Last but not the least, I express my deepest appreciation to my wife, Uikyoung Shin, who prayed for me and supported me for my study and health every day and without her sacrifice I wouldn't be where I am today.